ALSO BY MELANIE WARNER

Pandora's Lunchbox

The
MAGIC
FEATHER
EFFECT

*The Science of Alternative Medicine
and the Surprising Power of Belief*

MELANIE WARNER

SCRIBNER

New York   London   Toronto   Sydney   New Delhi

Scribner
An Imprint of Simon & Schuster, Inc.
1230 Avenue of the Americas
New York, NY 10020

First Scribner trade paperback edition January 2020

SCRIBNER and design are registered trademarks of The Gale Group, Inc., used
under license by Simon & Schuster, Inc., the publisher of this work.

For information about special discounts for bulk purchases, please contact Simon
& Schuster Special Sales at 1-866-506-1949 or business@simonandschuster.com.

The Simon & Schuster Speakers Bureau can bring authors to your live event. For
more information or to book an event, contact the Simon & Schuster Speakers
Bureau at 1-866-248-3049 or visit our website at www.simonspeakers.com.

Interior design by Jill Putorti

Manufactured in the United States of America

10 9 8 7 6 5 4 3 2 1

Library of Congress Cataloging-in-Publication Data
https://lccn.loc.gov/2018287914

ISBN 978-1-5011-2149-4
ISBN 978-1-5011-2150-0 (pbk)
ISBN 978-1-5011-2151-7 (ebook)

*For Harry, my dad, whose curiosity for
alternative medicine I like to think lives on in these pages*

Dumbo, look! I have got it! The magic feather! Now you can fly!

—TIMOTHY Q. MOUSE

Contents

Donna's Eden

Under the influence of an energy healer

It's Friday night in Asheville and the banquet hall is nearly full. The room buzzes with a giddy hum as people gather in little communities along a neatly assembled perimeter of chairs. Middle-aged women are the best-represented group, but I spot a few men and some young people, too. A few even look like teenagers. Above us, three chandeliers disperse a warm glow over the room.

I make my way past the tables scattered with paraphernalia about the weekend's "Introduction to Energy Medicine" and notice that, to everyone's delight, Donna Eden, the woman we have all come to see, has made an impromptu appearance on the stage. She darts from end to end, seemingly to retrieve something. Then she turns to the crowd, gives an enormous smile, and hoists her arms up over her head in a two-handed wave. The room erupts in applause.

When the program starts for real some fifteen minutes later, there is no introduction. Donna simply marches onto the stage with her husband, David Feinstein, who was once a psychologist on the faculty at Johns Hopkins University School of Medicine and is now an enthusiastic proponent of "energy psychology." David takes a seat at the back of the stage. Donna glides to the front, arms outstretched.

"Hello, everyone!" she says, beaming. "I welcome you all sooo much. I'm so glad you came. *I'm* so glad I came. I feel like there's a part of Asheville that's home for me. My mom was born here.

"I love teaching this. The world needs some really good healing tools that we can all use and depend on, and these are simple. They look like magic sometimes, but they work."

Donna's head is framed by a halo of blond curls, and as she speaks, her face lights up with a perpetual smile. "Our bodies produce the most profound medicine inside us already. It's energy. Energy is the oldest medicine there is. It's the safest and most organic and it's free. We need a medicine that's free. We need tools to be able to heal ourselves."

David gets up from his chair and moves to the front of the stage. "Donna isn't just talking here," he chimes in. "She's living proof of how energy medicine works. Several weeks ago, she celebrated her seventy-second birthday."

Around me, I watch a few eyes widen at the thought that this spunky, effervescent woman onstage is one of the oldest people in the room.

"What I'm going to share with you this weekend are basic hands-on techniques you can use every day to give you more vitality and joy. These can clear energy blocks, help you feel younger, and relieve pain."

As Donna talks, shadowy streams of pedestrians glide past the windows, which are dark now apart from the starry blips of light from the chandeliers. Asheville's streets are like this, never crowded but always busy. Nestled into North Carolina's ancient Blue Ridge Mountains, the town is populated by a motley assortment of college students, artists, hippies, retirees from the Northeast, and those who want to go to "healing sanctuaries" and open stores selling "gifts for the soul." It is a fitting place for energy healing.

Energy healing is a nebulous category. It encompasses older practices such as Reiki and newer ones such as Therapeutic Touch, Donna Eden's Eden Energy Medicine, and many other approaches you may never have heard of, such as polarity therapy, BodyTalk, Matrix Energetics, and Quantum-Touch. The simplest way to think of it is as a practice in which adherents believe the human body possesses a subtle force that can be harnessed for its own rejuvenation and repair. Energy healers say they manipulate this force through some combination of light touch, moving their hands at a distance over their clients' bodies, and merely holding a thought or intention about how they want someone's energy to move.

Although this mystical energy has never been validated by science—by which I mean replicated, rigorous, controlled studies—I haven't come here as a debunker. I'm not interested so much in how Donna Eden and other alternative practitioners are wrong, but in how, against the odds, they may be right. I am trying to stay open to possibilities.

Donna looks out into the audience. "I'm going to need a volunteer. Most people understand all this a lot better if I demonstrate it. Over the course of the weekend, I hope to have a lot of you up here."

About thirty hands shoot up, and Donna picks a stocky woman named Penny. When Penny arrives onstage, Donna reaches affectionately for her hand.

"I'm going to use a biofeedback tool that's built into our bodies. It's called energy testing. Sometimes it's known as muscle testing, but we're not testing muscle strength; the strength of a muscle doesn't change from moment to moment. What changes is the muscle's ability to resist pressure based on the amount of energy flowing through it. We're testing energies and how they're flowing and where they're stopping."

Donna looks intently at Penny for a few seconds. "Okay, now put your arm out like this." She props Penny's arm up at a ninety-degree angle from her body and puts one hand on Penny's shoulder and the other on her outstretched arm. "Now I'm going to push down on your arm and you try to resist."

Donna gives Penny's wrist a firm downward push. Her arm doesn't move. "Nice and strong," Donna notes.

Then she walks backward so she's about ten feet away from Penny. "I'm going to trace your meridians."

Donna starts swooping her hands through the air as though she were conducting an experimental orchestra. She traces the invisible outlines of Penny's body, from her head down to her brown suede boots. Then she walks back to Penny for another test. Just as before, she puts one hand on Penny's shoulder, the other on her wrist. This time, Penny's arm falls like a lead balloon, as if she weren't resisting at all.

A look of surprise streaks across Penny's face while the crowd exhales in amazement. I've watched Donna and others do this online before, but I still don't have a reasonable explanation of what's going on. Donna never appears to alter her level of effort, and Penny didn't look as if she were intentionally caving. Nor does it feel like a cheap setup. Donna radiates such enthusiasm and certitude that it's hard to conclude she is actively making any of this up or doesn't believe 1,000 percent in what she is telling us.

"What I did," Donna says, "is trace a meridian backwards. That takes energy out of the flow. Every single person in here has healing in their hands. I've never met anybody in my whole entire life who couldn't use their hands to move energy."

I shift about in my seat, suddenly aware of just how distant is the parallel universe I have entered. Meridians, a concept that originally hails from the traditional Chinese practice of acupuncture, are said to be invisible lines within the body along which energy flows. Each meridian is named after an organ—liver meridian, lung meridian, gallbladder meridian, etc. But just like healing energy, scientists have never pinpointed or otherwise identified such a thing inside the human body, much less figured out how someone might trace meridians backward or use them therapeutically. Donna doesn't typically do onstage healings, but in her 1998 book, *Energy Medicine*, she describes ridding people of all kinds of emotional and physical afflictions during the twenty years she spent doing one-on-one sessions. On her Eden Energy Medicine website, the fourteen hundred people she has trained as practitioners of her method also tell of often dramatic, life-changing improvements in their patients' migraines, arthritis, back pain, fibromyalgia, autoimmune skin disorders, and sleep, fertility, and digestive problems. Many of these accounts include a testimonial of profuse gratitude from the person who was once suffering. Donna attributes her recovery from multiple sclerosis in her early thirties to moving energy in her body.

When I began telling people I was writing a book on alternative medicine, I couldn't avoid accounts like this. People would tell me that they had been helped by some unconventional approach or that they knew someone who had. "Everyone has a story to tell," my husband marveled one day after a string of conversations with colleagues inquiring about his wife's daily activities.

There is reason to be skeptical of healing claims. As critics of alternative medicine rightly point out, people get better for many mundane reasons. Sometimes medical problems spontaneously and inexplicably disappear or improve on their own—the "treat a cold, it will last a week; leave it alone, it will be gone in seven days" principle. Such natural healing happens with many viral and bacterial infections, and from time to time with chronic pain and even cancer, particularly of the kidney, brain, lymphatic system, and skin. In other cases, symptoms wax and wane, particularly those such as pain, fatigue, and stiffness. Since we tend to seek help when things are awful, the chances of feeling better in the days after some particular treatment can be high. It's also easy to overlook that some diseases have natural quiet periods when symptoms temporarily disappear or improve. If you do a particular therapy just before this period of remission, it's normal to assign credit to it. Finally, people sometimes don't actually have the disease they've been diagnosed with. According to a 2014 study, diagnostic errors happen to one in twenty Americans.

But can we dismiss all such healing claims as misattribution or wishful thinking? As I look around at the rapt faces watching Donna, I can't help wonder about the way belief can sometimes create healing. In 1997, Arthur Shapiro, a noted psychiatrist, wrote, "Until very recently, the history of medical treatment was largely the history of the placebo effect." By this he meant that throughout history many popular medical treatments—crab eyes, lozenges made from dried vipers, pigeon and turtle blood, horse dung, and bloodletting to name a mere few—surely did not work as advertised, but did have occasional efficacy due to the placebo effect, the phenomenon whereby our expectations about a treatment's effectiveness can actually make it effective. Just as the generously eared elephant Dumbo believed he could fly after his trusted mouse friend, Timothy, gave him an ordinary crow's feather to carry in his trunk, telling him it was magic, it seems the human body has ways of eradicating illness based entirely on what's in our minds. Once called the "coolest strangest thing in medicine" by the British doctor and writer Ben Goldacre, the placebo effect has been the subject of much eager research in recent years.

In any given year, across the US 19 million people see a chiropractor, 3 million do acupuncture, 3 million practice the slow, mindful exercises of tai chi or qigong, and 1 million visit an energy healer such as Donna. At least 24 million do meditation or breathing exercises, which also haven't yet met the standards of evidence necessary to make them mainstream medical treatments. The popularity of alternative approaches—which can be defined as treatments not taught widely in medical schools, not broadly available in hospitals, or not possessing a clear scientific basis—is due to a variety of factors.* Most commonly people are seeking natural, noninvasive, side-effect-free ways to ease the symptoms of health problems that doctors can't cure and don't always have satisfying treatments for.

Proponents of these therapies all have their theories about how they work, and for this book I will spend time hearing them. But since I regard science to be, as the astrophysicist Carl Sagan put it, "by far the most successful claim to knowledge accessible to humans" and the best safeguard we have against our biases, I am more interested in what the surprising number of placebo researchers, neuroscientists, and psychologists studying alternative healing techniques or mind-body interactions have to say. What are the scientific reasons that seemingly unscientific practices might work? In this inquiry, I do not attempt to be comprehensive. Rather than catalog the vast universe of alternative treatments, I focus on a subset of those with the great-

---

*A word about terminology: Although the terms *complementary medicine* or *integrative medicine* are more accurate in that most people do not use these therapies to the exclusion of standard medicine, I've chosen to speak about *alternative medicine* because it's a more recognizable term.

est potential to elicit placebo effects and potentially other types of mind-body healing.

Donna calls for more volunteers and does additional demonstrations in which she studies someone for several moments and then does something to either weaken or strengthen the person's energy patterns, which she says she can see. To "disrupt the energies" of a tall middle-aged man named Karl, she flutters her fingers over several points on his chest and arms. As she pushes on his arm, he strains mightily to prevent a seventy-two-year old woman from overpowering him in front of two hundred people. The energies of an elderly man in a wheelchair, which initially test strong, are described as becoming weak when he reads from a textbook. "Half your meridians run in one direction, and half run in the other," Donna explains. This man's energies, she says, are flowing in a way that makes it difficult for him to perform mental activities such as reading. Donna leads him in an exercise she says is designed to shift his energies back into balance. When he reads from the book again, his arm tests strong.

I stare at the elderly man and try to imagine the meridians in his chest stopping in midstream and reversing course, either thwarting or restoring some internal zest. But I can't seem to get past the stripes on his sweater, which, when I glare at them, blend together in a hazy blob. By the time Donna dismisses us for the evening, I realize I am ready for a break.

The next morning when I arrive at the event hall, Donna's daughter Titanya is waking everyone up with an "energy dance," with loud Middle Eastern music and lots of arm and hip swinging. I try to join in but quickly realize I can't dance and drink coffee at the same time, and I opt for the energy swirling in my cup.

Donna appears, fully awake and raring to go, and tells us that we will be learning exercises to get our energies moving freely and in the right direction, to give us more of that vitality, joy, and pain relief she mentioned yesterday. These exercises consist of rubbing, holding, or tapping various points on our bodies, many of them the same ones used by acupuncturists. But instead of poking thin needles into the skin, Donna uses hands. She has us tap with both sets of fingers on our "K27" kidney meridian point, located on either side of the chest just below the collarbone. This is intended to deliver oomph when we are tired and to keep the immune system strong amid stress. A notch below that, in the middle of the chest, is the thymus gland. We thump our fingers on that. "Also excellent for your immune system," Donna chirps. We rub spots just below our rib cage—our so-called spleen meridian. All I can feel is my actual spleen, which isn't sure about being kneaded. At one point, I glance around the room and see two hundred people hitting themselves in the face. Of course, I'm one of them.

We also team up with a neighbor to practice. Donna says we should gently press on certain points on our partners—their hands, their forehead, their feet—to see if we can get their energies to move, testing it before and after with the arm push. If their arm proves weak, we are supposed to do something to see if we can strengthen it, and vice versa. I partner up with Charity, a twentysomething from Georgia, who drags her fingers around my face and head to stimulate my gallbladder meridian. Later, I clutch her two naked big toes in an attempt to create deep relaxation and promote the release of toxins. This is called a Brazilian toe technique.

None of it seems to be working. Charity is a wisp of a woman and I am five feet ten. Every time I test her she is weak. Her arm flops right down. I know it isn't a fair setup, but Donna has said this isn't supposed to be about strength. So either I'm hopeless at fortifying Charity's energies or she has some serious deficiencies.

Donna tells those of us having trouble to take it slow and try practicing at home with some simple exercises, such as tapping on our cheekbones, which "grounds" a person by sending energy down the body and out the feet. "You become more sensitized over time and then can eventually feel the actual flow of energies in your body," she says. "When I hit my cheekbones, I can feel energy going down my legs, but it doesn't start like this."

This is just one of many of Donna's experiences. In her book, she explains that ever since she was a little girl, she's seen multiple bands of color around and within other people. Some of them had texture and shape; others moved and vibrated. For a long time she didn't think this was strange and assumed others saw such colors, too. Her mother, who also said she saw them, encouraged her daughter's perceptions. Donna writes that when people are physically and emotionally healthy, their energies have a harmonious flow, appearing as "an endless waterfall," spilling over the top of the head and caressing the body. With illness, energy looks blocked, frenzied, or chaotic, like static on a TV screen (back when TVs had static).

The next day, we see Donna only briefly. She leaves us with more inspiring messages about our ability to heal ourselves and take control of our health with our energies, though not necessarily, she is careful to point out, at the expense of conventional medicine. "Modern medicine is wonderful and has its role," she notes. The rest of the morning is given over to Donna's husband, David, who does a presentation on energy psychology, which uses tapping on various points and meridians to try to cure phobias or past traumas. I duck out early to meet Donna, finding her in the eleventh-floor hallway of her hotel, saying good-bye to one of the event's organizers. "Ooohhh, I was looking for you at the book signings," she says, referring to the breaks during which she signed copies of *Energy Medicine.* "I thought, 'Where is that journalist Melanie?' Did you get your book signed?" Then a big laugh and the enormous, infectious smile.

Donna wraps her hand around mine and pulls me into her hotel suite, where she is frantically packing for her next trip. Just a few days ago, she was in England, and tonight she and David leave for Arizona. I tell her that my copy of her book is digital, which left me with nothing to sign.

"Oh, that explains it. Well, I like to use that time to really get to know people, and I sign their books in different colors depending on what I see for their life color, which is the outermost layer of the aura."

This, I realize, is why a giant pile of markers was sprawled on the table in every imaginable color. Of course, I want to hear my life color anyway.

"Well, first of all, you're violet. That's really interesting to me. I don't know what your life has been about or anything. . . ." She stops and looks at the window behind the couch I'm sitting in. "Can I trade places with you? There's just so much energy from outside. I just want to look at you."

We swap couches. "So the aura has all these auric fields or bands that go around a person, and I can always see at least seven layers out. There's so much information in all of them, but there's one band that never changes. You're born with and you die with a certain color, and you're a violet. You are soooo violet. When I saw you coming in the hallway, I thought, 'Oh, God, here comes a violet!'" She laughs. "And then I saw your yellow sweater and thought, 'It's so complementary with your purple.'

"Violets are really smart, but I always think that in past lifetimes they have either been the victims of the world or ruled the world, and when they got to the end of their lives, they said, 'This is it?' So this is the lifetime to crash through that. They never reached that spirituality or what they hungered for. There's always introspection, but their challenge is to surrender and let go and have some fun. My husband is a violet life color. More left-brained; they rely on their intellects. So that's you."

Setting aside that I don't believe in past lives, I think it's not a bad read. I have on occasion been told that I need to lighten up and not take things so seriously, which always annoys me, probably proving the point. For some reason, hearing Donna tell me this isn't the least bit irksome. Her way of conveying information is disarmingly modest and devoid of any judgment. It's as if she just *knows* it.

Donna says she is a blue life color. "Blues are all about healing other people, they're caregivers. A lot of nurses are blues. They also don't have a lot of critical thinking. They just want to believe in the positive. So it's not like I want to see the positive, I actually see it. That pure place. Even the darkest person, at the soul level, has a beautiful place of purity that is stronger than evil if it's fostered. Connecting with people at that pure energetic core happens in the deepest healing work."

"So if a guy sleeps with his best friend's wife, you don't think, 'What a douche bag'?" I ask.

"I don't. And it's gotten me into trouble because even if I see really wretched things in their energy, I tend to look deeper to see their pain and their struggles, and I connect with them at that level. Witnessing them in this deeper way brings out their natural goodness and somehow sparks their healing. I love engaging people at this soul level."

I must have knitted my eyebrows in confusion.

"Here, I'll tell you what it looks like." She describes a man who was standing outside a bodega near her house in California "looking really puzzled." Donna went into the store but couldn't help looking at him through the store's open front. "His etheric field, the band closest to his body, looked totally smashed in. That would mean he did not have any good reflections of himself, no real mirror for himself. I thought, 'What a hard world for him to have that out of commission.'"

The second band in the man's aura looked "flimsy," and the third showed that he was what Donna calls a kinesthetic, someone who easily senses and takes on the feelings of others. "This would make it even harder for him because he felt it all. He didn't have any kind of hard shell from feeling everything or from worrying for others. If you just looked at his face, you might have thought, 'Maybe I shouldn't park my car there.' You might think he was dangerous and someone to stay away from. The truth is he probably has had a really hard life. His auric field was just decimated."

Donna walked out of the store, went up the man, and handed him a bag of avocados. "They gave me these, but I really don't need them. Would you like them?" she asked.

Suddenly, Donna says, there was a different man. "Not only did his whole face light up, his auric field brightened and expanded." It wasn't the avocados, but the spirit of empathy in which they were given. "We human beings need one another. We heal one another just by recogniz-

ing each other." This ability to pick other people's emotional cues—
or "energies"—was of immense value during Donna's two decades of
one-on-one practice with patients. Today, most of her time is spent
on the road teaching and doing presentations such as this weekend's,
but she says that her unusually vivid perceptions enabled her to form
hunches about what was driving someone's illness and what might be
done to help relieve it. Sometimes she saw past or current relationships causing emotional stress. Other times there would be a deep
sadness or personal failure. Some people, like the California man, were
crushed and overwhelmed by their abundant feelings. Often people
were profoundly fearful of their illness or other problems in their life,
and Donna focused on soothing their panic.

Before I leave her warm, affirming orbit, I ask Donna about the
arm pushing and energy testing she did so many times onstage. The
practice is widely used in alternative medicine. Many homeopaths
and naturopaths employ it to test people for food allergies or to see
which particular homeopathic remedy to choose. Some chiropractors also use it. In his book *Serve to Win*, Serbian tennis star Novak
Djokovic says he found out he was sensitive to gluten after his nutri-
tionist told him to hold a piece of bread against his stomach and then
was able to push down on his weakened arm. Although this energy or
muscle testing is alleged to reveal something factual about a person,
nearly every time scientists subject it to properly controlled conditions, the method falls apart. In one of the more recent examples,
three experienced chiropractors pushed on the outstretched arms of
fifty-one patients while they were holding either a vial of saline solution, which was presumed to have no effect on someone's energies,
or a toxic compound, which was agreed would weaken one's energy.
Neither the chiropractors nor the subjects knew which vial contained
which. In 151 of these blinded iterations, the toxic substance was
identified correctly only 80 times, or about the rate you'd get from
guessing. The chiropractors, in other words, were wrong nearly as
often as they were right. And this wasn't research done by a skeptic.
The lead investigator told me that he is a believer in remote viewing,

which is what we're now calling ESP. He would have loved to verify energy testing as a "nonlocal" phenomenon.

I mention all this to Donna, who reacts with excitement and no hint of defensiveness. "It's so interesting—can I talk about that? I totally understand why some of the studies wouldn't get it. There are several things."

This leads into several threads of conversation—about a group of curious doctors she taught in the nineties and a far less friendly group in the seventies that sued her for practicing without a license (she eventually got a license as a massage therapist since there are no licenses for energy healing)—before she remembers I had a question. "What were you asking about again?"

"Energy testing studies."

"Oh, right. People often think they're so intuitive when they get into any kind of healing, so you have the thought in your mind that this one is going to be strong, this one is going to be weak, and then you transfer that energy to the other person. You've got to have your mind completely out of the way."

It takes me a moment to realize that Donna is saying that energy testing can be wrong because the tester may unintentionally be altering the other person's body in some way.

"Isn't that a bit far-fetched?" I ask, picturing some kind of Jedi mind trick.

"Here, I'll just show you."

Donna moves next to me on the couch and takes hold of my right arm, propping it up at the ninety-degree angle I'd seen her do so many times onstage. She pushes down on my wrist. My arm wiggles against her force but doesn't move.

"Okay, so you're nice and strong. I'm going to do something."

The next few seconds are quick and undramatic. She closes her eyes, then opens them, takes my arm, and pushes down. This time my arm collapses as if it were raw chicken. I am stunned to realize that I have no ability to prevent this. As my arm is dropping, I am thinking that I will at any moment locate the strength to halt the fall.

"I didn't test you with more pressure," she says, before I can even ask. I can't say I felt more pressure either. "I just thought, 'Oh, she's going to be weak.'"

I walk out of Donna's hotel with an itchy, unsettled feeling. Despite my skepticism about healing energy, I find myself catching glimpses of the appeal of Donna's worldview. Maybe it's her perpetually cheery disposition or the way she seems to size up someone's spirit by reading their "energy fields," but I find myself trusting that she somehow knows exactly what she's doing. I start to realize that in the right hands, belief can become contagious, especially when it contains the promise of relief from physical suffering. As William Osler, one of the founders of modern medicine, wrote in 1904, "Faith in the gods or in the saints cures one, faith in little pills another, hypnotic suggestion a third, faith in a plain common doctor a fourth. In all ages . . . the mental attitude of the supplicant seems to be of more consequence than the powers to which the prayer is addressed." But how far could such magic feathers take you? If you believe a person can help heal your body by moving energy, flipping your meridians, clearing your chakras (the seven centers of spiritual energy along the midline of the body), or putting thin needles into carefully designated points on your body, is there any scientific evidence that this can become a self-fulfilling prophecy? And for what kind of ailments? I have no idea, but that's what we're going to find out, starting with one of the world's oldest surviving medical practices.

Lost in Translation

The enduring practice of acupuncture

In March of 2015, George O'Maille lay languidly on a reclining chair with his eyes closed. It was the last hour of a four-hour class to help people alleviate or better manage their pain, and George was listening to the soft-toned instructions of a guided meditation. "Imagine choosing what you want to let through that narrow passageway of nerves in your neck," a therapist said, as she slowly walked around the room twirling a rainstick. "Think of what it would be like if you could flip a switch and stop those signals causing pain, giving yourself more mobility and comfort in your body." As the rainstick rumbled, an acupuncturist came by and poked thin needles into George's ears. They pinched a little going in, but George didn't mind. He had far more worrisome sensations to attend to.

The whole morning he had been thinking about how he should leave the class early to go home and take some of his pain pills. In a scramble to leave the house he'd forgotten to, and his distress was starting to build. Six years earlier, George, a sixty-nine-year-old former production manager for a television station in Sacramento, had recovered from a bout of shingles only to be left with excruciating pain, a complication that occurs in about 20 percent of cases. This postherpetic neuralgia, as it is officially termed, is characterized by burning or shooting sensations and thought to be associated with nerve damage. For George, it had settled into his right side. His chest, armpit, shoulder, and elbow often felt, as he drily put it, as if someone "took a baseball bat, stuck nails out of it, dipped it in tar, lit it on fire, and then beat me with it." All of the opioids and anti-inflammatories his doctors prescribed and the steroid shots they injected into his spine barely affected the pain. What finally helped was gabapentin, an antiseizure medication often prescribed for this condition. George took it religiously three times a day, helping to dull, though not eliminate, his agony.

Now that the class was nearly over, he tried to set these thoughts aside. As he lay there, his body like a jar of warm honey and three needles sticking out of each ear, George noticed something shift. "I felt the pain going away and going away and going away, and I thought, 'This is really strange,'" he recalls. By the end of the session, the pain was completely gone, and it stayed away even though he stopped taking gabapentin. After about a month, a few of the deep aches and sharp stabs returned. But instead of reaching into the medicine cabinet, George got another acupuncture treatment, this one an individual session. "And, boom, it was gone again." It's been six months, and he says he is again starting to feel a little pain, but is planning to go for another acupuncture "tune-up."

I'm hearing this story in the office of Dr. Daniel Neides, the former director of the Cleveland Clinic's Center for Integrative & Lifestyle Medicine, the place where George took his pain class and got his acupuncture treatments. Dr. Neides is trying to convince me that not only does acupuncture work wonders, it does so because of how the needles manipulate energy—or what acupuncturists call qi (an indispensable Scrabble word pronounced *chee*)*—running along the body's twelve meridian lines. Like Donna, he maintains these conduits are real. In this sense, you could call acupuncture a kind of

---

*Sometimes this word is written as *chi* or *ch'i*, which was a compromise between English and German pronunciations. *Qi*, however, is the transliteration from Chinese and is considered the more accurate spelling.

energy healing, although owing to its long history of use in Asia, no one does. Acupuncture gets its own designation.

"Now you can't argue that," he is saying about George. "At some point you have to go, 'I gotta believe because it's just happening time and time again.' I see people improve all the time."

"But this doesn't necessarily mean meridians exist," I object. "Aren't there other reasons why people might improve after acupuncture treatments?"

"The Chinese have been working with this for thousands of years. There's just no question about it. It has to do with meridians." Neides is wearing his white doctor's coat and has dark hair and a boyish face. He is getting increasingly worked up as he talks. "I could prove to you that you have an energy force right now. Follow what I'm doing." He puts his hands together and rubs them vigorously up and down. "Keep rubbing hard. Create the friction. Now take your hands about an inch or two apart like this and tell me you don't feel like there's something in between your hands."

Something does seem to hover between my hands. When I slowly move them apart, it feels as though I am pulling against a warm, airy, tingling glob. The feeling is not particularly dramatic, but when I concentrate on it, it's there. I mention to Neides that scientists might nominate the heat creation and nerve activation from friction as a possible explanation.

"That's positive and negative charges. If you ask me how this works, it's that you're manipulating positive and negative ions in a way that can reduce inflammation."

I'm not sure what this means, but I don't press the point because I don't yet know enough to argue my way out of what feels as if it's going to be a dense forest of pseudoscientific mumbo jumbo. (In 2017, Dr. Neides left the Cleveland Clinic after an uproar over an antivaccine blog post he wrote.) I also don't raise the possibility that George's pain spontaneously disappeared on its own because the timing of such a thing after six years seems particularly uncanny. Plus, how could there have been another extremely coincidental instance of pain relief a few months later? "What about a placebo effect?" I say weakly.

"I don't know any placebo effect that could do that. Do you?" Neides says, his eyebrows arched.

I'm not sure I do. Although I know a placebo effect is a beneficial response to a belief or expectation about a treatment, I don't know how complete or long-term it can be, how quickly or slowly it takes effect, or what conditions it might help alleviate. When people talk about the phenomenon, it is usually with an air of vague dismissiveness, as if placebo effects were fleeting, banal things that happen to other people, not to you or someone you're telling some incredible story about. "Just a placebo," they might say.

Neides tells me about some of the other things that happen here at the Center for Integrative & Lifestyle Medicine, which is on the second floor of a large building about ten miles east of the Cleveland Clinic's main glossy sprawl of hospital facilities and doctors' offices. In addition to acupuncture and pain-management classes, patients come here for a variety of other nonmainstream services such as Reiki, yoga, meditation, massage, and holistic psychotherapy. (While yoga, meditation, and massage are hardly fringe these days and no one would question their ability to lift moods or tone shoulder muscles, when it comes to treating disease, they're alternative.) The five primary-care doctors at the center, most of them MDs, run a suite of nonstandard tests designed to look for stress, inflammation, dietary issues, and gut-bacteria imbalances.

"We have hundreds of people on a wait list to get in to see the doctors," Neides boasts. "In 2010, we had four hundred patient visits the whole year. In 2015, we had twenty-six thousand. People are looking for something different, and we really have a whole different mindset here."

Before coming to the center in 2014, Dr. Neides worked for fifteen years as a family doctor in Cleveland, where he routinely spent less than twenty minutes with each patient. It got to a point, he says, where he wasn't sure how much he was helping his patients. When he was approached to run the center, he leaped at the chance. Now his patient appointments and those of the other doctors run at least double the amount of time they used to, which allows him to have more substantial conversations about what he and others call life-style medicine—focusing on nutrition, exercise, sleep, and stress.

That the Cleveland Clinic, which consistently rates as one of America's top hospitals, has a dedicated, well-appointed center offering such things as Reiki and acupuncture demonstrates how firmly alternative practices have gotten their foot inside the door of mainstream medicine. When the Cleveland Clinic's center first started in 2008, its handful of doctors and therapists saw patients inside dingy small offices in one of the hospital's oldest buildings. Now the place, located in the Cleveland suburbs, practically has its own zip code. On the way in, I wound my way around a road that was surrounded by sixty-eight verdant acres of fields, forests, and horse trails, all of it donated to the Cleveland Clinic in 2002 by the defense contractor TRW during its merger with Northrop Grumman (which had earlier purchased the sprawling property from an oil heiress and congress-woman). And it's hardly just the Cleveland Clinic. Many major hospitals and universities around the country have similar, albeit less generously situated, centers. The Mayo Clinic, MD Anderson Cancer Center, Massachusetts General Hospital, Harvard University, Johns Hopkins Medicine, Stanford Health Care, and the University of Maryland School of Medicine all run some kind of "complementary" or "integrative" facility that patients can go to on their own or that their doctors refer them to.

After I leave Neides's office, I wander down the hall to find Jared West, the acupuncturist who put those needles into George's ear. I'm hoping he'll be able to enlighten me on how a person's pain could suddenly and unexpectedly dissolve from a treatment such as this. West is tall, with a soft, meditative voice. He's been involved with traditional Chinese medicine since his childhood, during which he learned acupuncture from his mother. He also practices qigong and tai chi, which are related methods of slow, focused movements synced with one's breathing. Both hail from China; qigong is the older, broader practice and tai chi is a subset of it developed perhaps in the seventeenth century. Tai chi, which is short for *t'ai chi ch'üan*, was originally a martial art and a lesser-known form of it still incorporates the more rapid, self-defense maneuvers. In the US and Europe, the most popular variety is the palliative, non-martial-art kind. West learned qigong and tai chi at the age of eleven when he traveled to China with a Taoist priest.

Sitting in his cubicle outside a row of treatment rooms, West tells me that the ear points he used that day on George and the other seven pain-class participants are helpful for "calming the nervous system and changing the way it responds to pain." Although George's experience might have been more dramatic than most, West says he witnesses acupuncture as effective for pain all the time. According to traditional Chinese medicine, many health problems, including pain, are the result of a deficiency or blockage in the flow of qi or blood along some particular meridian line. By inserting and gently twisting the needles in points along those lines, you can help stimulate the flows of qi and blood, bringing it to needed areas and thus alleviating the pain.

Although most of the meridians are named after and travel through organs, West explains that they span the whole body, often traveling far beyond their namesake. This is why, he says, needles often don't go where you think they should—why, for example, "supporting the lungs" could mean putting needles into the shoulder, arms, and hands, or why treating "liver qi stagnation" is not done anywhere near the stomach but through points on the feet and hands. Similarly, a treatment for headaches might also find an acupuncturist sticking needles into the feet, the logic being that headaches are sometimes caused by excessive qi in the head. Thus needles into the far reaches of the body will redistribute this energy and remove the stagnation. The ear points West used are believed to balance the distribution of qi through the entire body.

The few times I got acupuncture, years ago from a friend who had just graduated from acupuncture school, I never thought too much about how it worked. My friend would occasionally mention qi and its movements, but I was mostly focused on the sensation of being poked with needles, which often felt bothersome, as if I had a fly on my skin that needed to be flicked off. Invariably though, I would forget about this mild annoyance and dissolve into a relaxed, soupy state.

"Chinese medicine is a completely different system of understanding how the body operates," West says. "It's about functional relationships and a dynamic view of the body, not isolated body parts. Its effects can be quantified by conventional medicine, but its mechanisms are harder to explain with what we know about anatomy and physiology."

This appears to be an understatement. West explains that five natural elements influence the movement of qi in the body—wood, fire, earth, metal, and water. Imbalance in the wood element, for instance, connotes rigidity and tightness. "Someone with a woody pulse might be tighter either physically in their tendons or emotionally with an inclination to yell and get frustrated by things not going right in their life," he says. Back pain in an older person could be the result of too much "cold" (absence of fire element) and applying heat would relieve it. There are also the opposing forces of yin and yang. Migraine headaches, in addition to too much head qi, could be attributable to the failure of the liver meridian to hold yin energy, such that yang rises and causes a headache.

West urges me not to think of these things too literally, and I can tell he is trying to be patient with my obviously Western brain, but his explanations are sounding to me more like weather reports than medical evaluations. I can't make heads or tails of phrases like "woody pulse" and statements such as "Spring is ruled by the wood element, which is associated with the liver," something I read in a book called *Between Heaven and Earth: A Guide to Chinese Medicine*, give me a headache. When I attempt to apply the basics of anatomy and physiology to these ideas, it's like mixing up the contents of two puzzles and trying to put them together. But why are these two medical systems so different and why isn't there overlap? What I need to do, I realize, is lead my search back in time to the places that gave rise to these two divergent worldviews.

The history of what we call Western medicine is well known. Its roots, as any medical student can tell you, lie in the sea-encircled lands of ancient Greece some twenty-five hundred years ago. There, doctors such as Hippocrates began the first stages of systematic, scientific thinking about the human body. For centuries prior, people in Greece and nearly everywhere else on the planet believed that illness was the result of some supernatural force that had gotten ahold of you. A demon had cursed you or a god was extracting punishment for some offending behavior. The era's doctors were shamans, and their method of healing used rituals to appease such angry spirits or cast them out. They would chant, make sacrifices, and use remedies made from plants and animals. Sometimes they would enter the spirit world through trance states that were psychedelically induced.

Hippocrates, Galen, and other Greek doctors rejected this supernatural approach. They believed natural laws governed the operations of the body, and they wanted to figure out what they were, which was about as grandiose an ambition as you could imagine for the time, on par with trying to figure out what made the sun shine or the sky blue. Outside of the identification of major organs from corpse dissection, virtually nothing was known. How body parts worked together to generate not just health and disease, but life itself, was a complete mystery. No one had any idea that the heart is a pump, that signals from the brain move muscles, and that traits from our parents are passed down through DNA. All of this sophisticated knowledge was some two thousand years away.

The system the Greeks came up with was based on bodily fluids, or "essential humors." There were four—blood (secreted by the liver), phlegm (secreted by the lungs), and yellow and black bile (emerging from the gallbladder and spleen, respectively)—and they governed not just one's physical health but also temperament and emotions. If you were low energy, depressed, or highly sensitive to your surroundings, you had an overabundance of phlegm, an idea that endures in the word *phlegmatic*. If you had anger issues, were bold and ambitious and quick to envy, you had too much yellow bile. Black bile, on the other hand, made one pensive, melancholy, withdrawn, and overly cautious. A balance of the passions, Hippocrates wrote, was fundamental to good physical health. Each humor was mapped to a specific organ and one of the four natural elements—earth, water, air, and fire.

Since maintaining good health required keeping the humors in balance, the Greeks were constantly trying to coax one or more of these fluids out of the body. Bloodletting was a common practice for releasing an overabundance of blood. For the other humors, there was liberal use of enemas, laxatives, saunas, purgatives, and teas that made you sweat. Many of these scatological traditions continued in Europe until, shockingly, the nineteenth century. In the eighteenth century, Voltaire advised young men to find wives who could administer enemas pleasantly and quickly, which I think should provide women everywhere with a deep sense of gratitude for having not been born in eighteenth-century France. Bloodletting too endured as a mainstream medical treatment, until the 1850s, and not just for the poor. In 1799, well-meaning doctors took eighty ounces of blood from President George Washington's arm in his last twelve hours on earth, probably hastening his death from a throat infection. Doctors' lancets were so symbolically important that one of the world's earliest, and now one of its most influential, medical journals was named the *Lancet*.

Several hundred years later and five thousand miles away from ancient Greece, a similar revolution in human thinking was going on in ancient China. This one, though, did not trickle down to the wider populace at the time and was instead undertaken by a relatively small number of elites. No one knows much about who these people were, where they lived, how they got their ideas, or even how many of them there were. What is known is that sometime in the second and first centuries BC, they wrote texts that were collected into what's now considered the foundational source for traditional Chinese medicine and the first written account of acupuncture—the *Huangdi Neijing* (Yellow Emperor's inner canon).

Much like the Greeks, these Chinese authors were turning away from the heavens. They too had decided disease was not caused by gods and demons but by innate systems inside the body. In their view, two such phenomena existed—blood, or *xuě*, and what they called qi. In the *Neijing*, qi has numerous meanings, including the invisible force that surrounds the earth so it doesn't fall into the sun and the force regulating wind, moisture, heat, cold, and other aspects of nature. In the context of the human body, there were two connotations. Much like the Greeks, who called the breath *pneuma*, the ancient Chinese decided something about breathing was profoundly life-giving, and they called this indispensable essence qi. Qi also referred to the life-sustaining properties of food. The character's literal translation, I am surprised to learn, is "steam rising from rice."

"Qi has nothing to do with energy," Paul Unschuld sniffs over the phone one morning from his office in Berlin. "That idea is pure nonsense." Unschuld is a Chinese medical scholar who ran what was then called the Institute for the History of Medicine at Ludwig Maximilian University in Munich for twenty years and is now at Berlin's Charité-Medical University. I called him up because he has written a dozen books on the history of Chinese medicine and is the first Western scholar to have translated both books of the *Huangdi Neijing* into English, a painstaking endeavor that consumed the better part of the last forty years. It took such a long time, in part, he says, because he refused to transpose Chinese characters into words or concepts that didn't exist at the time, such as *pathogen*, *cell*, and, interestingly, *energy* and *meridian*.

Unschuld says that in the *Neijing* both blood and qi are described as traveling through the body in a deeply embedded system of hose-like "channel vessels." Some of the book's authors say that blood and qi flow in opposite directions in the same channels. Others, perhaps reflecting different traditions from different regions in China, maintain that blood and qi have their own separate channels. Referred to in the texts as *jing luo* and *jing mai*, these pathways, Unschuld says, have been completely mistranslated as "meridians." They were blood vessels.

According to Unschuld, the Chinese believed, much like the Greeks, that you had to keep the flow of essential bodily substances in balance. Too much or too little blood or qi in one area of the body got you into trouble. Instead of teas, saunas, enemas, and purgatives, the Chinese devised a sophisticated system for inserting needles into points along the *jing luo* and *jing mai* channels to draw out a combination of blood and qi. Sometimes this was with "hair fine needles" possibly similar to what acupuncturists use today (minus, of course, the sterilization and packaging). Other times, it was with more hostile devices, perhaps akin to what the Greeks employed. "There are nine different types of needles described in the texts, and many of these were nothing other than miniature weapons: mini-daggers, mini-swords, mini-lancets," Unschuld says. One variety was a fierce seven inches long.

"It's very explicit; it's not interpretation," he says. "The treatment with needles means 'open the blood vessel and with the blood, the disease will vanish.' It was used for a lot of diseases. It's always tied to terms like *remove*. Something has to be removed and the easiest way is with blood because you can see it. Qi you can't see, you have to imagine it."

In other words, ancient acupuncture was part modern needling and part bloodletting, though probably not the pints of blood the Greeks and later European doctors were known to draw. Unschuld says it's unclear how or when things shifted away from bloodletting and toward a gore-free version of needling, as there's no written record of such a transition. Texts written in the sixteenth century, though, focus mostly on qi manipulation, not bloodletting.

The influence of Greek humoral theories on the practice of Western medicine waned in the nineteenth and early twentieth centuries amid the advent of scientific experimentation and testing. Discoveries about cells, biological systems, and disease-causing bacteria revealed humors and bloodletting to be entirely wrong ideas about how the body operated, and they were eventually cast aside to the point where there is no trace of them in the current practice of medicine (though Hippocrates's prescriptions on medical ethics, exercise, fresh air, rest, and a good diet continue to ring true).

In China, the advance of modern scientific knowledge took a very different path. In an attempt to move China forward, the Daoguang emperor issued an imperial edict in 1822 that banned acupuncture and moxibustion, which is the practice of burning herbs along acupuncture points, from the Imperial Medical Academy. But China lacked academic institutions that could provide a home for medical science and help shepherd a widespread cultural acceptance. Instead, both ancient practices were kept alive by folk healers throughout China until the 1960s, when Chairman Mao systematically revived them in a bid to boost Chinese nationalism and provide cheap health care to the nation's sprawling population. From there, acupuncture made its way to the US in the 1970s as Americans started traveling to the People's Republic.

But before that, some Europeans made the journey to China and discovered the exotic practice of acupuncture, which had always involved more grace and art than the Greek-inspired traditions. In Unschuld's view, one of these Europeans is responsible for the current misunderstandings of acupuncture's origins. "George Soulié de Morant," he says, his voice dripping with rancor. "That French charlatan. He introduced these two erroneous concepts, qi as energy and the *jing luo* as meridians."

Unschuld's villain published the first popular book outside of China on acupuncture in 1939. George Soulié de Morant didn't have any formal training in medicine or Chinese languages, but harbored an abiding interest in Chinese culture and an ability to pick up languages quickly. When he moved to Shanghai in 1901 as a clerk for a French bank, he reportedly witnessed surgeries done using acupuncture as the only anesthetic. Soulié de Morant also claimed to have observed healings from acupuncture, including during a cholera epidemic in Beijing in 1908. Upon returning to France in 1910, he sought to introduce this fascinating Eastern practice to the West and found French physicians willing to learn it. He eventually wrote the book *L'acuponcture chinoise*, in which he summoned the French geographical word *meridien* to interpret *jing luo* and *jing mai*, and translated *qi* as "energy."

"Both these concepts," Unschuld says, "were disseminated in schools in France before people like myself could look into the ancient texts and see what the literal translation and meaning of these terms was. So now, most literature published in Western languages on Chinese medicine reflects Western expectations rather than Chinese historical reality." Morant may have misrepresented other things, too. According to research done by Hanjo Lehmann, another German scholar of traditional Chinese medicine, Morant wasn't a nobleman as his name implies; he added the *Soulié de* part to boost his status. Nor was he a diplomat in the French embassy, as he claimed. He worked as a translator. And there is no record of a cholera outbreak in Beijing or anywhere else in China around 1908, much less of acupuncture being able to help to treat it.

Lorraine Wilcox, a licensed acupuncturist who lives in California, holds similar views of Morant's translation errors, yet has a more generous view of his intentions: "I think he was trying not to give the Western doctors he was teaching too much of a culture shock. He wanted to make it sound more scientific and acceptable. He was trying to bring Chinese medicine to the doctors rather than bring the Western doctors to Chinese medicine."

As for why the Chinese haven't pointed out that qi doesn't mean "energy" and that meridians are channel vessels that originally had much to do with blood vessels, Unschuld says it's an issue of both translation and pride. In the 1970s, when acupuncture was becoming widely known in the West, few Chinese speakers had any idea how to properly interpret the complex concept of qi into other languages. So they went along with what English speakers had come up with. Also, thanks to Mao, traditional Chinese medicine has become something of a cultural export for China, and those devoted to its preservation aren't eager to do anything that might devalue or disrupt it.

Before talking to Unschuld, I had gotten the idea that the notion of qi running through meridians must have originated with some meditating Chinese monks. I imagined them retiring to mountain caves thousands of years ago, where they achieved enlightened or otherwise heightened perceptions of their bodily sensations. These monks then mapped, my story went, all the things that came to be known as meridians and acupuncture points. I don't know how I got this half-baked narrative, but in my defense I don't think I completely fabricated it. Several acupuncture websites I poked around on suggested as much.

In considering acupuncture's ancient origins, I had succumbed to thinking that because this practice has endured for so long, its founders must have been onto something that we here in the modern West have yet to quite put our fingers on. Some spiritual essence of the universe within the body perhaps, or an aspect of our consciousness connected to all things. Traditional Chinese medicine *does* have a holistic appeal, with its interconnected view of the body's systems and an emphasis on keeping these systems in harmony so as to prevent disease, as well as a deep understanding of the multiple, interlocking causes of disease—concepts that many Western doctors often fail to fully grasp or appreciate. Over the last two hundred some years, it's safe to say that Western medicine has been far too willing to myopically separate the body into component parts and systems.

But I now know that the writers of the *Huangdi Neijing* weren't meditating mystics summoning spiritual secrets. There probably weren't even any mountaintops. As some of history's early rationalists, they were undertaking a rudimentary and prescientific search for the factors that support life—the oxygen in the air we breathe that travels to every cell in the body, and the nutrients, vitamins, and minerals in food. These concepts are now so unremarkable and embedded into our worldview that it's hard to imagine what it must have been like not to know them, to wonder what the relationships between air, food, and being alive were. But now that years of careful, rational, bias-controlled experimentation and analysis have afforded us such knowledge, what's the point of talking about qi as a real thing flowing inside our bodies? Why adopt the second century BC view? It's a bit like arguing for the ancient Greek view that the liver secretes blood. When it comes to science, there is no East and West. Done properly and rigorously, there is just science. The anatomical and physiological underpinnings of "Western medicine" are scientific, while most of the two-thousand-year-old concepts of traditional Chinese medicine aren't.

But this isn't the same thing as saying acupuncture is useless. A few decades ago, some researchers who weren't acupuncturists used the scientific method to see whether this durable practice could reliably help people and what might be underpinning it. Their findings lead us headfirst into a consideration of what it means for the mind to play a role in healing.

Telltale Toothpicks

Acupuncture and the randomized,
placebo-controlled trial

In the mid-nineties in Seattle, Daniel Cherkin had a problem. All around him, it seemed, there was backlash. Flannel-shirted, punk-lite musicians were taking up arms against the boppy music of the previous decade. A hip coffee culture was recoiling from the watery crap the rest of America was drinking every morning. In medicine, people were becoming increasingly uneasy with accepting only what their doctors had to offer and were seeking help from acupuncturists, chiropractors, and massage therapists. Throughout Washington State, these practitioners garnered enough support to successfully lobby the legislature for a law requiring every insurance company operating in the state to cover their services. Passed in 1995, the law, the first of its kind in the country, mandated insurance coverage for any provider licensed to practice in Washington State, which meant not just MDs, physical therapists, dentists, and the like, but also acupuncturists, chiropractors, naturopathic physicians, osteopathic physicians, and massage therapists. (To become licensed, a profession must get regulatory laws passed at the state level.)

Then a senior scientific investigator for the research division of Group Health, one of Washington's largest insurance companies, Daniel Cherkin was supposed to conduct research on which treatments could actually help people. Group Health, now part of Kaiser Permanente, prided itself on not wasting money for things that didn't have solid scientific evidence, whether they were high-tech procedures or ancient practices. But the new law meant the company might have to pay for treatments that were giant question marks, such as acupuncture. "A lot of people who were interested in studying acupuncture up until around the year 2000 wanted to show it worked," Cherkin tells me. Most of these studies, he says, were small and had spotty methodologies.

So Cherkin decided to do a large, well-designed acupuncture trial to see whether it was effective for chronic pain, which is the problem that drives the most people to try it and is a symptom acupuncturists insist the practice is effective at treating. Officially, chronic pain is defined as bodily pain that lasts for more than three months, which is the general time frame of normal bodily healing. With an undergraduate statistics degree and PhD in epidemiology, Cherkin's interest wasn't personal. He had never been to an acupuncturist and knew little of the theories behind it. Yet he was interested in trying it. Just prior to the start of the study, to help him write the patient consent form, he got his first and only treatment, which he describes as "mildly annoying."

Because he worked at a health care company, Cherkin was able to round up a large number of people (638) to participate in the study. He selected patients with low-back pain because this condition was—and still is—both the most common type of chronic pain and the leading health complaint driving Americans to seek alternative treatments. According to US government data, 32 million people suffer from frequent or ongoing back pain and the 17 percent of them who seek alternative treatments do so because the stuff doctors promote—surgery, injections, anti-inflammatories, and opioids—often aren't sufficiently helpful, especially given their potential complications and side effects. Studies have shown that between 20 and 40 percent of those going under the knife to help back pain end up with what is officially and dispiritingly called "failed back surgery syndrome," meaning they're probably worse off than they were before.

"Most people are coming in to doctors with back pain and being told they have back pain, which they already knew," Cherkin says. "They're not getting a clear diagnosis or explanation because the docs are not well trained in this. People are frustrated, both the docs and the patients."

To see if acupuncture could be of any greater benefit than standard treatments, Cherkin divided his 638 patients—all of whom had had pain for at least three months and were new to acupuncture—into four different groups. One batch got seven weeks of standardized acupuncture treatments where needles were stuck into eight designated points on the body for a strictly prescribed amount of time. Another group got individualized treatments where the acupuncturists were free to work with as many needles as they wanted in as many points as they saw fit. A third group got a fake form of the therapy in which they got poked at non-acupuncture points with a toothpick encased in a skinny plastic tube. Since everyone in the study got the treatments while blindfolded and had never before had acupuncture, most participants couldn't figure out which group they were in. A final control group got no acupuncture at all and simply continued with their drugs, physical therapy, and doctor visits, which the other groups were free to continue with as well.

The results, published in the *Archives of Internal Medicine* in 2009, were revealing. None of the three different acupuncture treatments was any more effective than the others in reducing people's pain and disability. Patients who were poked with nonpenetrating kitchen toothpicks did just as well as those being properly needled in spots carefully selected by their acupuncturist. It seemed that all the fuss acupuncturists made about inserting needles just so into special acupuncture points along specific meridians was much ado about nothing. You could get the same effect with random pokes from toothpicks in your kitchen.

But all of these acupuncture treatments, if you can call them that, worked better than what patients were already getting. Roughly 60 percent of those who got acupuncture, or what they thought was acupuncture, saw improvements after eight weeks, as opposed to 39 percent of those in the group getting their usual treatment routine. This does not mean that acupuncture helped 60 percent of people or that 39 percent improved from their usual care. Some minority of patients would have improved anyway from random fluctuations in symptoms, and others might have been giving researchers overly rosy assessments of their symptoms, in what's known as a response bias. They could also have felt better because they were under observation in a clinical trial, which then inspired them to take better care of themselves—the Hawthorne effect. The 60 versus 39 percent was the relevant difference between the groups. Acupuncture had given people a measure of relief they weren't getting from their usual treatments from doctors, and it did so without the insertion of needles into special acupuncture points along certain meridians.

Similar conclusions were emerging around the same time in studies funded by German insurance companies. In other large acupuncture trials for headaches, knee osteoarthritis, and back pain, German researchers found that real acupuncture treatments weren't superior to the fake ones (they used blunt metal needles that poked the skin at non-acupuncture points and then silently retracted into little plastic tubes). Yet just as in Cherkin's study, the pain relief people got from both types of treatment were meaningfully better than those from usual care or being on a waiting list, and these improvements held for at least six months.

Aldous Huxley once wrote, "That a needle stuck into one's foot should improve the functioning of one's liver is obviously incredible . . . in terms of currently accepted physiological theory 'it makes no sense.'" Huxley, who was a believer in acupuncture, then added that maybe current theories needed to be modified. These studies suggested no urgency to rewrite theories on the liver or any other organ. Acupuncture's usefulness for pain, one of the German groups posited, was likely not due to special activity of the needles, but the manifestation of a "superplacebo."* Pain relief came from the mean-

---

*Confusingly, the phrase *placebo effect* has several meanings. For the placebo group of a clinical trial, it refers to any artifacts of natural and spontaneous healing and of the trial itself, such as the patients' tendency to want to please researchers with positive reports of improvement. This, as you might gather, is not what we're talking about here. The placebo effects that inspire and fascinate, and that this book explores, are those capable of creating real improvement.

ing people had assigned to their experience of acupuncture, or what they thought was acupuncture, and the beliefs and feelings the experience engendered. "[Our findings] force us to question the underlying action mechanism of acupuncture and lead us to ask whether the emphasis placed on learning the traditional Chinese acupuncture points may be superfluous," they wrote. In a study that pooled together the results from many clinical trials, the effects of acupuncture for chronic pain appeared more modest (rather than "super"), amounting to a 10 point reduction in someone's pain rating (on a 1–100 scale) as compared with the usual care. This reduction might not register much for someone with mild discomfort but would for those with excruciating pain. Another, more recent, roundup of studies, a meta-analysis, put the point difference at 30.

These findings left Cherkin and the German authors with a difficult question: Could you consider a placebo—whether a big one or modest one—a legitimate medical treatment? Acupuncture is by far the most studied alternative practice (surprisingly, more than a thousand clinical trials have been done), but there is evidence that other approaches might also work as placebos. Studies of chiropractic and osteopathic spinal manipulations show that these treatments can be of modest help with back pain, yet some show that a sham manipulation in which the practitioner is doing it wrong can also be of benefit. The same is true with homeopathy, which is not an herbal or natural concoction as is sometimes assumed, but the very definition of nothing. Homeopathic remedies are so diluted with water that it's like putting a raindrop into the ocean. The theories behind the practice center on the outlandish idea that water has "memory," and so, amid extreme dilution, there is still an active substance. Skeptics have thought to ask why it is that water, a long-lasting and promiscuous substance, doesn't also remember every other random, nasty, and infectious molecule it's picked up along the way.

Nevertheless, a 2005 review of 110 studies of homeopathy concluded that large, high-quality trials showed clinical effects for certain conditions, but no more than patients would have gotten taking a sugar pill. Some preliminary findings show that Reiki and other forms of gentle-touch energy healing such as Healing Touch can alleviate the stress, anxiety, and fatigue of cancer patients more than whatever usual approach they might be getting. Eden Energy Medicine, unfortunately, hasn't been subjected to any controlled studies, so we can't say with any certainty if it's an effective placebo.

Traditionally, for medical treatments to be considered effective, they have to be more than placebos. That is, they must beat a placebo treatment in the sort of randomized, blinded, placebo-controlled clinical trial that Cherkin and the German researchers did. This setup has been medicine's gold standard for the last fifty years and has led to massive advances in the treatment of disease. One of the reasons alternative practices are considered alternative is because they haven't been shown to meet this benchmark of being superior to their version of a placebo treatment.

In the years since that 2009 acupuncture study, Cherkin has spent many hours ruminating over this issue. "Do we dismiss acupuncture because we don't know for sure whether the effects are specific or nonspecific? Or do we say this is a valid treatment option which is safe, as long as it's done correctly, and there is evidence of effectiveness for some patients?" After much consideration, Cherkin, who is now retired, chose the latter conclusion. If patients were getting relief, who was he to judge? "I think we've totally lost sight of the importance of the healing encounter. If our goal is to do something that actually helps a person heal, then I think we need to consider a broader range of things as legitimate, especially when we lack other very effective treatments."

Others have come to this same conclusion. Following those half a dozen German acupuncture studies, health authorities in Germany decided in 2007 that acupuncture for chronic low-back pain and osteoarthritis of the knee, even though it's probably a placebo, should be included as a routine reimbursement by the country's social health insurance funds. A 2011 Institute of Medicine report called *Relieving Pain in America* also determined that using placebos wasn't a bad idea. "Placebos conceivably could be considered a form of treatment of pain, especially in light of the shortcomings of other modalities or other benefits they bring in their own right," the report's nineteen authors stated. But this is hardly everyone's conviction. After seven years of recommending that health care providers consider acupuncture for patients with low-back pain, the UK's National Institute for Health and Care Excellence reversed course in 2016 and stated that, because there's no evidence that acupuncture is more effective than a sham treatment, providers should not offer it for the condition. (It still recommends it for chronic tension headaches and migraines.)

In Cherkin's view, the "better than placebo" standard is problematic because it was not devised to test practices such as acupuncture, chiropractic, energy healing, tai chi, or meditation. The concept came of age during the boom years of drug development, when we needed to know what new remedies worked as advertised and were worth the inevitable risk of side effects. Figuring this out meant giving one group of people a sugar pill without their knowing that's what they were getting. This allowed you to peel off the riffraff of both bias and expectations from the important molecular effects of a drug.

But how does one fake meditate? Or do sham tai chi without realizing it's not the real thing? It's even debated now whether retractable needles (or toothpicks) are an inert treatment for acupuncture. Chiropractors and other manual therapists similarly struggle to find their sugar pill.* "For a lot of nondrug treatments, it's not clear what an appropriate placebo would be," Cherkin says. "If you're a researcher and the funding agency expects you to have a credible placebo control and there isn't one, you're kind of in a bind." The psychotherapy profession, he notes, has been wrestling with this dilemma for years. With nondrug therapies, you also can't do the double blinding that's demanded by medicine's gold standard. While the patients getting the treatment (and the researchers collecting

---

*Although placebo pills were once made with sugar, today microcrystalline cellulose or the mineral talc is preferred.

data about it) can be in the dark, the psychiatrist, acupuncturist, or chiropractor administering it obviously can't.

But if we are to take the placebo effect seriously, we first need to know what it is we're dealing with. Could it have eradicated George O'Maille's years of postshingles neuropathic pain simply due to certain beneficial beliefs and states of mind he was coaxed into? If so, no one else was so lured that day at the Cleveland Clinic. Everyone had pain, but George was the only one to see it dissolve. Maybe, though, that's because neuropathic pain is known to be quite malleable to placebo treatments.

One afternoon, I flip through the book *You Are the Placebo* by a former chiropractor named Joe Dispenza. In the book and elsewhere, Dispenza tells the story of getting hit by an SUV during a triathlon and severely injuring the middle and lower parts of his back. Although his doctors strongly recommended surgery to implant stabilizing rods along his spinal column, Dispenza declined all this in favor of self-healing at home, which he says entailed spending every day intentionally connecting with the "intelligence, invisible consciousness" inside him. This, he says, allowed his "greater mind with unlimited power" to do the healing. After twelve weeks, Dispenza claims he got up and started living his life again, almost as if nothing had happened.

Now a busy member of the self-help healing circuit, Dispenza writes of people coming to his workshops and getting healed of cancer, multiple sclerosis, degenerative bone disease, and Hashimoto's disease, an autoimmune disorder of the thyroid gland. When you read the book, you get the impression that a powerful, unconstrained healing force inhabits all of us. To unleash it, all we have to do is get out of our own way, eradicating the negative, self-limiting views that prevent our body's inner intelligence from taking over.

There are many books like this, taking the placebo effect and mind-body healing in general and propelling us straight into miracle territory. In his 2009 book *Reinventing the Body, Resurrecting the Soul*, Deepak Chopra tells us the mind can help mend the physical body of any affliction because, at its core, our flesh is a "fiction" and every cell is made up of "two invisible ingredients: awareness and energy." Healing disease is a matter of directing awareness and moving energy. In the film *The Secret*, we are told that skin cancer can be avoided if we think positive thoughts, because "the thoughts we think actually create a vibration in the [body's] water that has memory."

To me, it all sounded far too good to be true. I decided it was time to see how the placebo effect actually worked.

The Pharmacy Within

What does a placebo effect feel like?

"Are you ready for pain stimulations?"

Luana Colloca is peering at me through shiny wireless glasses. She is short with dark hair, a heavy Italian accent, and a scholarly air.

"Of course," I say. But the truth is I don't feel ready at all. When I spoke with Colloca on the phone a few months ago, going to Baltimore to see if I was a placebo responder seemed like an excellent idea. Of course I wanted to subject myself to her pain devices and see what would happen. But now that I'm here, I realize I have no idea what kind of torment we're talking about. Will I require first aid afterward? Will there be a scar? The uncertainty itself is a source of pain.

For the past fifteen years, Colloca, a neuroscientist, has inflicted "pain stimulations" on countless willing participants, coupling it with healthy doses of purposeful lying. It all started after medical school, while she was getting her PhD in neuroscience at the University of Turin, many miles from the small town in southern Italy where she grew up. One day she was in an operating room where a Parkinson's patient was getting neuron-stimulating electrodes implanted in his brain as a treatment for his disease. She watched as her mentor, the Italian placebo researcher Fabrizio Benedetti, tracked and recorded the effects of a placebo treatment on this man's individual brain neurons. (Remarkably, patients are conscious during this deep brain stimulation surgery.) To witness in real time how a belief, in this case the patient's faith that an injection of saline solution was in fact a Parkinson's drug, could actually move brain cells blew Colloca away. In that moment, she knew her future lay in studying the elusive secrets of placebos. After spending nearly a decade in Benedetti's lab, she left Italy in 2010 to strike out on her own, taking a job studying placebos at the National Center for Complementary and Integrative Health, a unit of the National Institutes of Health, in Bethesda, Maryland. She's now at the University of Maryland in Baltimore.

I follow Colloca out of her small office and down a fluorescent-lit linoleum hallway. She stops at a large beige metal door and opens it, revealing an equally beige, windowless room about the size of a large walk-in closet. There is nothing to see here except a large black laptop with an alarming number of cords snaking from its sides. One of them is thick like a jumper cable.

I sit down in the large armchair facing the computer while Colloca tinkers with the tentacled device. A grad student appears to fasten a small black box that's attached to the end of the jumper cable to my left forearm. As she secures it with bright green tape, Colloca explains what is about to happen: She will send varying amounts of heat to the black box on my arm, and each time, I am to rate the sensation from one to ten according to the pain I feel. One is no pain. Ten is molten-lava hot. We'll do three rounds of ten blasts each, with breaks in between.

Colloca fetches another laptop and wheels it in front of me. "Every time I send heat," she says, "you will see either a green or red screen. The green will be low pain and the red will be high. After each one you will do a rating." The grad student fastens a mouse-type device to my right hand for this purpose.

It sounds simple enough, and when we get started, things are just as Colloca said they would be. Every time I see green, there is barely any pain, just a pleasant warm sensation. I rate it between one and three. When the screen is red, my skin starts screaming and I yearn for the sensations to stop. This might not be exactly what a hot iron feels like, but that's the image that jumps to mind, and I'm relieved when the pain lasts only a few seconds. I rate these between seven and nine. Maybe I'm trying to be tough, but I never award a ten.

Toward the middle of the second round, I start wondering what Colloca is up to. There's a ruse here somewhere, and my hunch is that Colloca is giving me the same levels of heat each time but making me think they're different because of the varying screen colors.

I turn out to be almost right. When it's over, Colloca tells me what happened. The first two rounds of heat were exactly as advertised. The red screen was paired with a 118°F heat signal, and the green screen with a mellow 80°F signal. This, plus the (false) prompts at the outset, Colloca explains, got my brain settled into the idea of the red-green arrangement, which I then took confidently into the last round, when every single one of the heat signals was full bore, at the upper limit of what the machine was capable of.* "I was not gentle with you," she says, grinning mischievously.

I take a minute to consider the bizarreness of this. What had first felt scalding hot was then like sliding into a warm bubble bath. The pain simply wasn't there, almost as if the fiery 118° signal had disappeared. Except it hadn't. Anyone who doubts that the placebo effect is real might want to go sit in Colloca's pain chair. All it took for my pain to vanish was the simple belief that I wouldn't feel pain because of a green screen. As soon as the red color came back, so did my belief that there would be a fierce blast of heat. But the signal never changed, only some silent circuitry of faith inside my head. The placebo effect had snuck up on me, and I could suddenly see

---

*Another word about terminology: The question of how to distinguish the brain from the mind is arguably the most pressing unsolved question of neuroscience. For the most part, I use *brain* in the context of scientific research on this organ or neural activity that is largely unconscious or outside of our control. *Mind* is applied to the complex matrix of feelings, thoughts, beliefs, memories, and stories that we are aware of and can, at least theoretically, gain access to. No doubt it's an imperfect linguistic system, mostly because the distinction is woefully incomplete. Nobody can say how or where all the aspects of a mind live inside a brain, or how (or whether) brain activity gives rise to consciousness.

how such a mystifying experience, to the extent that it would occur in the real world and not just in labs—and why wouldn't it?—could easily be misinterpreted as some supernatural phenomenon. Under different circumstances, someone might have claimed I'd done some mind-over-matter abracadabra to lower the heat with my thoughts or to block it from traveling up my arm using my energy. "It didn't feel like a placebo," a woman had told me at a party a few months earlier. She'd done acupuncture for back pain, and her discomfort had resolved right after the treatment. But how could anyone know what a placebo felt like or be able to pinpoint when it was occurring?

Colloca explains that what I experienced was a result of the human brain's remarkable ability to become conditioned to patterns. As predictive, pattern-hunting machines, our brains are continuously assessing the millions of things happening in our environment at any moment and trying to make sense of it all. Is this person trustworthy? Is it safe to walk down this street? What does it mean that every time I wash my car it rains? Connecting the dots of past experience and present cues is the essence of how we function and survive in the world. It also makes us highly susceptible to suggestion. Sometimes we see patterns that aren't there or we stitch together stories that aren't true, much as I did with the red-green screens. But in the case of a placebo effect, belief becomes its own kind of truth. If your brain has concluded something is going to take your pain away, it just might.

Even more fascinating is that my attempts to think my way out of this had no effect whatsoever. Trying to anticipate and piece together how Colloca was deceiving me didn't erase the magic. My brain was off and running on its own, and I was powerless to stop it, just as I hadn't consciously created the effect in the first place.

"You are a high placebo responder," Colloca remarks. "It's always so interesting when people who know about placebos or don't think they are going to be responders turn out to be. We still want to believe that if a person guesses that all the stimuli is at a high level, that that will make a difference. But it just doesn't."

She says that roughly 60 percent of the people she brings into her lab and tests in this fashion, with both the verbal suggestion and conditioning rounds, are placebo responders like me. A great many of us can block our pain and relieve other symptoms without ever realizing it.

No one knows exactly when, how, or why placebo effects came into existence, but chances are these hidden skills have been with humanity for quite some time. The ancient-Greek physician and philosopher Galen must have been observing some versions of the phenomena when he marveled that sick people at the temple of the healing god Asclepius were occasionally cured by nothing more than a "shock to the mind." Much later, in 1807, Thomas Jefferson wrote favorably of the perpetration of "pious fraud" by doctors: "One of the most successful physicians I have ever known has assured me that he used more bread pills, drops of coloured water, and powders of hickory ashes, than of all other medicines put together."

The word *placebo* is said to have originated with the Catholic mass that priests performed for the dead in the first millennium. At these services, mourners chanted the ninth verse of Psalm 116. But because the Church's official Latin version of the Hebrew Bible included a mistranslation, a slightly different version was sung. Instead of "I will walk before the Lord in the land of the living," it was "I shall *please* the Lord in the land of the living": *Placebo Domino in regione vivorum.* By the thirteenth century the funeral mass itself became known as the placebo and featured insincere mourners who were either paid to be there or came for the magnanimous feast that followed. In medicine, the first known use of the word was in a 1772 lecture by the influential Scottish physician and chemist William Cullen, who told of administering a patient a dose of mustard powder: "I did not trust much to it, but I gave it because it is necessary to give a medicine, and as what I call a placebo."

Not until the middle of the twentieth century, when an anesthesiologist noticed something peculiar about his wounded patients, did anyone endeavor to study placebo effects. Henry Beecher was a professor at Harvard who joined the US Army during World War II and served as a doctor on battlefields in Italy. There, in makeshift military field hospitals, soldiers arrived with serious injuries—broken bones jutting though the skin, lacerations in the arms and legs, and what Beecher described as "penetrating wounds" of the thorax, abdomen, and head. Yet, despite their trauma, three-quarters of these men said they had no pain or only slight to moderate pain.

At first, Beecher figured this was because they were in shock. Although that was the case for some, most were deemed mentally alert. These men simply weren't feeling their horrific wounds. Beecher was astonished. At home in Boston, where he'd worked as the chief anesthesiologist at Massachusetts General Hospital, he'd seen civilians with similarly severe injuries from car accidents and other violent trauma display far greater levels of distress. He began compiling data on his soldiers and then, when the war was over, compared it to that on injured civilians in Boston. Publishing his findings in 1955, he revealed that 83 percent of injured civilians requested pain relief upon their arrival at the hospital, whereas on the battlefield only 32 percent of soldiers did.

Pain, Beecher concluded, wasn't entirely what it seemed. He began to consider the implications of this for the painkillers he doled out to patients. Did some kind of internal mechanism turn off or tamp down pain without drugs? Beecher had to consider the possibility that anytime he gave painkillers, part of the resulting relief was not from the drugs themselves, but from the patients' conviction that they were going to feel better. Perhaps, Beecher reasoned, the effects of drugs such as morphine, papaverine, and codeine were being overestimated.

But why did so many more soldiers experience this effect than civilians? Beecher had a theory. He reasoned that to a battlefield soldier, the hospital signified an extraction from "an exceedingly dangerous environment, one filled with fatigue, discomfort, anxiety, fear and real danger of death. The hospital is 'a ticket to safety.' . . . His troubles are about over, or he thinks they are." Although a wounded soldier might ultimately lose a limb or the ability to walk, he wasn't going to be returning home to his family in a flag-wrapped box. He would probably be welcomed as a hero. For civilians, however, hospitals held a very different meaning. Admission marked "the beginning of disaster." A civilian might worry how he was going to pay for the inevitable onslaught of treatments or when he was going to be able to get back to work. Far from being a safe refuge, the hospital represented a menacing threshold between a predictable existence and a challenging life of uncertainty.

At Harvard, Beecher set up the first program to study placebo effects and in 1955 published his most famous paper on the subject, "The Powerful Placebo." After collecting fifteen drug studies that had compared fake pills to the real thing, he concluded that placebos relieved the various symptoms (not only pain) of between 26 and 58 percent of 1,082 patients, which he then averaged into the neat, widely publicized number of 35 percent.

Beecher's finding that a treatment with no active pharmacological ingredients could help fully a third of all patients got a great deal of attention and helped catapult placebos to an official place in medicine. Not until many years later did someone point out the grave errors Beecher had made, ones that would go on to muddy the waters of placebo research for years. The Harvard anesthesiologist had overlooked that a 35 percent response rate in the placebo arm of a study did not mean that 35 percent of the people in that group had improved because of self-healing effects. Among those 1,082 patients, many had conditions known to resolve on their own or to come and go, such as headaches, postoperative wound pain, cough, and even the common cold. In 1997, a pair of German medical statisticians concluded that the natural history of illnesses was a major factor in ten of Beecher's fifteen studies. The only way you could have separated out the real placebo effect in these studies would have been to include a third set of patients who didn't get any treatments and were instead on a wait list for future treatment, which is an extra and unnecessary expense when testing the efficacy of drugs. Placebo researchers now realize that there is no neat number for the effectiveness of placebos, and that responses will vary widely depending on the condition being treated, how a placebo is delivered, and the beliefs of those being tested.

Still, in his examination of placebos Beecher did get a lot of things right. For example, he showed how important it was to assess drugs with placebo effects in mind. In 1970, the FDA mandated for the first time that drugmakers conduct "adequate and well-controlled clinical evaluations" of drugs they hoped to bring to market, emphasizing that a placebo comparison group was "desirable." Even though not formally required, the use of placebo groups became commonplace, thus rendering the randomized, double-blind, placebo-controlled trial the gold standard.

The placebo effect then faded into so much medical wallpaper. No one was particularly interested in studying how it worked or what it really was. To most researchers, a placebo effect was a nuisance phenomenon that you had to beat if you wanted to get a lucrative new drug onto the market.

But in the late 1970s, two neuroscientists at the University of California in San Francisco stumbled onto one of the biological reasons beliefs and expectations can lessen pain. Recently it had been discovered that not only were there receptors for opioid drugs such as morphine in the human brain, but that this is because our neurons make their own painkilling chemicals. Howard Fields and Jon Levine dug into this consequential finding with a series of experiments on dental patients who had just gotten their impacted wisdom teeth extracted. In one of these trials, they gave a group of patients an IV drip of saline solution and told them it was morphine. The patients received it immediately after they'd had their teeth yanked, and as a result, 40 percent of them experienced a reduction in or leveling off of their pain. But this manifestation of a placebo effect wasn't surprising. What excited Fields and Levine was that the drug naloxone, a chemical that blocks the effects of morphine and is routinely used to pull people out of drug overdoses, erased everyone's placebo effect.

The patients who started feeling better after getting saline saw a return of their pain when naloxone was put into their IV.

This suggested that a placebo for pain worked in a manner similar to morphine and used the same neural hardware. Instead of the drug attaching to our opioid receptors, the brain's own stash of neurotransmitters—chemicals that make our brains function by passing information across one neuronal synapse to the next—did the latching. Fields and Levine had shown for the first time that a placebo effect generated an objective biological response in the brain. A whole trove of painkilling, opioid-like molecules (endorphins and enkephalins, for instance) moved in response to a person's expectations.

No longer could you frame this as errors of bias or meaningless fluctuations. In follow-up experiments, Fields and Levine determined that just expecting to feel better—e.g., watching a saline solution that you thought was morphine go into your arm—could be as powerful as six to eight milligrams of morphine, which is a dose effective for many patients.

Then in the 1990s, it was discovered that additional brain neurotransmitters sprang into action during other types of placebo effects. Scientists at the University of British Columbia sent radioactive tracers into the brains of Parkinson's patients and found that for many of them the injection of saline solution they thought was a Parkinson's medication prompted the release of large amounts of dopamine, one of the brain's most important neurotransmitters. Among dopamine's many roles in the brain is the production of that coveted feeling of being rewarded, whether from watching an episode of *Stranger Things*, getting retweets, or drinking an after-work beer. Dopamine is released both when you anticipate doing these things and when you actually do them. During a placebo treatment, this chemical appears in response to your anticipation of something that is going to help you. Although it's not quite clear what role this plays in a placebo for pain or other symptoms, in Parkinson's patients additional dopamine is transformative since deficiencies in this substance are what cause the disease's tremors and movement difficulties. Drugs such as levodopa and apomorphine are effectively dopamine replacements. As, apparently, are saline injections or sugar pills. The Canadian scientists found that inactive treatments people believed were real boosted dopamine levels in Parkinson's patients up to 200 percent and temporarily restored normal movement.

The brain's endocannabinoids, which were discovered in the mid-1980s and attach to the same kind of receptors as do the active ingredients in marijuana, are also thought to play a role in a placebo effect for pain, although one different from those pain-blocking endorphins first identified by Fields and Levine. This finding hammers home that there is not one neat, unified placebo effect but many. Even when addressing a single symptom, different mechanisms may be driving a response, and what exactly many of these mechanisms are remains largely unknown. Outside of those studies on pain and Parkinson's disease, the inner neurological workings of placebos haven't been well investigated.

But what propels our brains to release such helpful doses of neurotransmitters? In 2008, a Harvard researcher did a particularly revealing study on the topic. Ted Kaptchuk had worked as an acupuncturist in Boston and over the years noticed that many of his patients would recover in ways he didn't think were attributable to acupuncture. After twenty years of practice, he concluded that a lot of what he'd been doing had nothing to do with needling. "I am a damn good healer," he told the *New Yorker* in 2011. ". . . If you needed help and you came to me, you would get better. Thousands of people have. Because, in the end, it isn't really about the needles. It's about the man."

Kaptchuk reasoned that it was less about the needles than one's abilities to elicit placebo effects while using them. What was it, he wondered, that made someone good at this? He closed up his acupuncture practice and became a researcher, getting himself positions at alternative medicine centers associated with Harvard Medical School. After writing a series of thought-provoking articles in major medical journals—bearing titles such as "Placebo Effects: The Dark

Side of Randomized Controlled Trials" and "The Persuasive Appeal of Alternative Medicine"—Kaptchuk started doing something no one had done before: randomized, controlled trials that specifically tested placebos, as opposed to just employing them as control groups for drugs or other treatments. For the 2008 study, he gathered 262 people who suffered from irritable bowel syndrome, a condition marked by abdominal pain and other gastrointestinal distress with no obvious cause. One set of patients got no treatment and were put on a wait list to serve as the control. Two other groups received acupuncture with retractable needles. One got three weeks of impersonal, perfunctory treatment with no touching by the acupuncturist beyond what was required to simulate insertion of the needles and no unnecessary talking or personal discussions. The other half got enhanced and highly empathetic sessions in which an acupuncturist asked questions about a patient's life and the difficulties their symptoms caused them. He or she said things like "I'm so glad to meet you" and "I know how difficult this is for you." The acupuncturist also touched the patient's hands or shoulders and spent at least twenty seconds lost in "thoughtful silence."

When the results were tallied, of the people who got the warm, empathetic placebo acupuncture, more said they felt "adequate relief" than did those in the impersonal group or on the wait list—62 percent versus 44 percent and 28 percent. Outcomes for the severity of people's symptoms and their perceived quality of life were similar. The idea that friendly, positive, attentive, and anxiety-reducing interactions could spur placebo effects was not entirely surprising. Earlier nonplacebo studies had shown that for certain non-life-threatening conditions simply having supportive and reassuring interactions with your doctor could lead to better outcomes than perfunctory, non-definitive, and conversationally challenged ones. But most of these studies were not randomized controlled trials or didn't set out to explicitly test the relationship between a patient and a practitioner. Kaptchuk was one of the first to clearly attribute symptom improvement to the way two people interacted with each other, thus showing what a successful placebo effect might look like in a clinical setting. It wasn't just a matter of what someone consciously expected or hoped would happen. There was also the way a doctor or a practitioner made you feel about the whole experience. Beliefs were intimately tied up with a far broader and more complex terrain of thoughts, feelings, and assessments. Harvard was sufficiently impressed with this and other studies to make Kaptchuk a director, in 2011, of a new Program in Placebo Studies, which remains his current role.

In one of Kaptchuk's earlier articles, he had noted that sessions with alternative practitioners such as acupuncturists and chiropractors often had an abundance of all the things he would later detail in the IBS study. Patients had relatively lengthy encounters with a person who made them feel supported and listened to and who gave them explanations of what exactly was wrong—which meridian was blocked or deficient, or which vertebra was out of alignment. Such precise and meaningful diagnoses, whether accurate or not, have been shown in studies to improve outcomes. "In unconventional medicine," Kaptchuk observed, "patient experience is never devalued or brushed aside as unreliable." Additionally, patients often had heightened expectations about healing: "An exaggerated notion of the possible readily elicits patients' magical anticipation." Much as the Germans had proposed acupuncture as a "superplacebo," Kaptchuk wondered if alternative treatments—with their hands-on, congenial atmosphere—might be "an especially successful placebo-generating health care system."

But Kaptchuk isn't the placebo researcher who can tell me more about this idea. In recent years, he has stopped talking publicly about most matters related to alternative medicine, due to the controversies it tends to stir up. "I have too many friends on both sides," he told me. He also reminded me that alternative medicine hardly holds a monopoly on "placebogenic potentials." Modern surgery, for example, is likely to be one of the most potent placebos there is. Studies have shown that undergoing vertebroplasty for spinal fractures, arthroscopy for osteoarthritic knees or swimmer's shoulder, and percutaneous coronary intervention for chest pain, in which a tiny balloon and stent are inserted into an obstructed artery, are no more effective than having a surgeon give a placebo incision. Just thinking you've gotten your shoulder or knee fixed is as good for pain relief and mobility as actually having the procedure done.

I first imagine the San Diego Pain Summit as a high-octane assembly of tanned and chiseled California CrossFit enthusiasts. I see them all in a large warehouse of sweat and tire rolling, drinking protein smoothies, on their way to looking like John Cena and Serena Williams. That the San Diego Pain Summit is really a gathering of friendly physical therapists, more interested in getting rid of pain than inflicting it, is only slightly disappointing. The low-key conference is held over two perfectly sunny days in a series of storm-worn, single-story buildings at the water's edge along San Diego's picturesque Mission Bay.

I've come here because Fabrizio Benedetti, the Italian neuroscientist and Luana Colloca's former boss, is scheduled to speak. Considered one of the world's preeminent placebo researchers, Benedetti has done more to uncover the neurobiological underpinnings of placebo effects than anyone else. He replicated and validated Fields and Levine's initial findings about endorphins and identified a role for endocannabinoids. He has also done extensive research on Parkinson's patients. Perhaps most impressively, he documented a placebo response within a single neuron of a patient's brain—the feat Luana Colloca said inspired her career, and which we should contemplate while noting that the human brain contains no less than 80 billion neurons.

A professor of neuroscience at the University of Turin Medical School, Benedetti seems a meticulous scientist who presents his data without rhapsodizing. During his San Diego presentation, his voice is steady as he cycles through footnoted slides about milestones in placebo research and shows brain scans of people undergoing a placebo effect. In one, we see increased activity in the rostral anterior cingulate cortex and the dorsolateral prefrontal cortex. In another, it's decreased in the dorsal anterior cingulate cortex, the insula, and the thalamus. Around the room, I notice blank expressions.

Heads perk up, though, when Benedetti plays a short video of his work with Parkinson's patients. We watch a man's hands shaking and fumbling as he tries to place checkers onto the black spots of a game board. The video cuts to the same hands after Benedetti has turned on the man's deep-brain stimulator, a set of electrodes implanted into the brain to stimulate certain neurons. With the device on, the man moves his hands smoothly to put the checkers onto the squares.

Then we see the same man after his stimulator has been turned off but without his knowledge. Thinking it's still on, he moves his hands as steadily and gracefully as if he were at a piano playing Chopin. A wave of amazement spreads over the room, and Benedetti explains that in his lab in Turin approximately half of the Parkinson's patients he tests this way are able to do what this man did, thanks to a belief and expectation-driven dopamine release.

When Benedetti finishes his presentation, a young man follows him outside. His eyes look wide and hungry as he explains that his father has Parkinson's. Quoting Benedetti's assertion that the effect also works with medication, he asks if such a thing could work for his father, who has been experiencing side effects from frequent medication doses. Could he take a sugar pill and get the same benefits?

Benedetti's long and detailed answer amounts to *No, your dad probably can't just take a sugar pill or do saline injections.* He explains that the placebo responses occurring in his lab, although strong, are short-lived, lasting roughly thirty minutes, as opposed to ninety minutes for the Parkinson's drug apomorphine. Since these drugs have to be taken so frequently, sometimes every few hours, it's unlikely that a sugar pill would hold up for that many treatments. Moreover, Benedetti tells him, from time to time people get the same side effects from placebos as they do from the drug. The bad expectations sometimes can't be divorced from the good ones. The young man still appears hopeful as he leaves. He tells Benedetti he is grateful for all the information. Yet it's unclear what, if anything, he will be able to do with it.

His question is much the same one that I'm wrestling with: What are those suffering from chronic diseases supposed to do given that their brains have a potentially formidable internal pharmacy?

"There's a difference between cure and care," Benedetti tells me as we head off once the clutch of supplicants disperses, "and I think that placebos are more about care than cure. You feel better about yourself, your condition. You have better quality of life, your symptoms aren't as bad, at least temporarily."

"Are you saying placebo effects are temporary?" I ask.

"The most honest answer is, we don't know. Because when we want to study the psychological and biological mechanisms, we have patients under strict control conditions, in the lab for six hours or twelve hours. We don't know what happens after six months, twelve months, or even after eight weeks."

Benedetti and I have wandered into an empty conference room where one lone table has yet to be folded away for the day. "Placebos don't work everywhere," he continues. "They are really effective in pain, anxiety, depression, and motor disorders like Parkinson's. But, at least as far as we know today, they don't work at all in conditions such as cancer and infectious diseases."

In 2003, French oncologists rounded up seven cancer studies in which a placebo group was included (for ethical reasons, cancer trials don't often include placebos) and found that only 2.7 percent of patients who got a placebo pill or fake chemotherapy infusions experienced a reduction in their tumors, which is close to what you would expect to see from cancer's natural progression.

There also isn't any evidence, Benedetti says, that expectation and belief-driven placebos can undo the effects of common conditions such as heart disease, diabetes, osteoporosis, autoimmune disorders, or Alzheimer's.* In fact, many people with Alzheimer's don't respond

---

*Some drug studies seem to show that placebos can heal ulcers, lower cholesterol, and reduce blood pressure, but it's hard to draw conclusions from these studies because none of them include a control group to account for fluctuations in natural healing or biases inherent to clinical trials.

to placebos at all because the degeneration of their brain's prefrontal cortex makes them unable to anticipate the future. A placebo effect is not, as the self-healed triathlete Joe Dispenza and others would have us believe, an unlimited healing force. There is a placebo effect for certain things and not others.

"I think one of the big future challenges for placebo research is to really understand in which medical conditions it works," Benedetti says.

I ask him about using placebos for things where we know there's a strong effect, such as pain. "Even if placebo effects are temporary, like most drugs, shouldn't we try and use them to help people?"

"If you're asking me if it's ethical to use a placebo in routine medical practice, my answer is no. Because if patients realize that you are deceiving them during the doctor/patient relationship, it's very bad for medicine and the medical community."

I tell Benedetti I was thinking more of alternative treatments and their "placebogenic potential."

Benedetti shifts uncomfortably in his chair. "I think that's a very dangerous way of thinking. It's true that the more complex the therapeutic ritual is, the more powerful its positive effects on the patient's expectations and the more positive the therapeutic outcome. But is it ethical to use everything and anything to induce positive expectations? Just imagine a quack performing a bizarre ritual. Is he justified to enhance the patient's expectations in this way? Like this table, this chair." He knocks on the table and motions to the metal chair underneath him. "Maybe they could be part of some ritual."

I find myself agreeing with him. "Like maybe you perform a ceremony with a chair over your head and some people might think it's so strange for a doctor to do this that it must be some kind of special healing."

"Yes. I could tell my patients this is a very powerful treatment for your headache. You can induce strong positive expectations for sure. Unfortunately, many quacks use this kind of approach. I get deluges from them—objects, talismans, bizarre procedures or rituals. They tell me, 'I have found a good way to increase expectations.'"

"But many alternative healers seem to genuinely believe in what they're doing. They don't appear to be trying to deceive anyone."

"What is the definition of a quack? It's a therapist using a treatment which is not recognized by mainstream medicine."

Technically, when I look it up later, it's a "fraudulent or ignorant pretender to a medical skill," but ignorant is close enough I suppose.

"Do you think an acupuncturist is a quack?" I ask. Outside the door, I hear the hushed mumble of people leaving the conference for the day and heading in the direction of free booze. Benedetti appears to be thinking.

"That's a good question," he says slowly. "But in my opinion, yes. Because acupuncture is not yet in mainstream and conventional medicine. There is no real study showing that acupuncture or Reiki or tai chi is more than a placebo."

Benedetti's response surprises me. This is not just because I had assumed that someone spending his career studying a concept would be more sympathetic to its use, but because, even amid its prescientific roots and tangle of meteorological mumbo jumbo, I would have thought *quackery* too harsh a term for acupuncture. Among alternative practitioners, acupuncturists enjoy the highest rate of US physician referral and are gainfully employed at many major American cancer centers to help patients with the stress, pain, anxiety, and other side effects of the disease and its treatments. If Benedetti thinks acupuncturists are quacks, I can only imagine his views of those who do Reiki or Eden Energy Medicine.

He raises a good point though. Where does the line for placebo treatments end? If acupuncture, chiropractic adjustments, and tai chi are marginally acceptable, does this swing the barn door open to a stampede of healing crystals, magnet beds, brain-wave devices, and a thousand other things you can buy on the internet, in supplement stores, or at holistic fairs? I suppose it could, although the ingredients for a "successful placebo-generating health care system" are probably going to include the kind of credibility and cultural recognition that only the most popular alternative treatments enjoy. And since the strongest placebo responses come from therapeutic rituals that are meaningful and intricate, the dizzying array of pills and supplements on the market are likely to have weaker effects unless they are coupled with visits to a practitioner, such as a naturopathic doctor (a licensed professional focusing heavily on herbs, supplements, and homeopathic remedies) or a homeopath.

Benedetti's other concern, about medical ethics, is an ongoing topic of debate among placebo researchers. Surveys reveal that patients do not welcome the idea of their doctors slipping them sugar pills. We want our doctors to stick to knowable facts and valid treatments, which can sometimes mean sending us off as worried and perplexed as we came in. Alternative practitioners can, in contrast, say pretty much whatever they choose and give any treatment or diagnosis they please. As Ted Kaptchuk observed, "Inevitably . . . the chiropractor will find the subluxation, the acupuncturist will detect the yin-yang disharmony. . . . They never fail to find a problem. By rooting pain in a clear physical cause, chiropractic validates the patient's experience." Such discrepancy in standards hardly seems fair and sets up an easy runway for fraud or complete nonsense.

But back when I was in Baltimore, I heard a very different view about alternative medicine from Benedetti's protégé Luana Colloca: "If a treatment has been shown to be effective or if an individual patient believes it is helping them—if it releases anxiety, improves the quality of life, or otherwise improves outcomes—then why not?"

By "shown to be effective" Colloca was referring to the wider parameter of being better than whatever usual care people were getting, instead of the "better than placebo" criterion Benedetti was demanding.

"But," she hastened to add, "I would love to see more honesty around it, too. Maybe you inform a patient and say we don't know why acupuncture or something else works, but we know there is a big placebo component."

Although it's tempting to think that such honesty might ruin a placebo effect, this is not necessarily the case. Two recent studies have shown that if people with chronic conditions (specifically irritable bowel syndrome and low-back pain) are given vials clearly marked PLACEBO and told that a placebo effect can be "powerful" and has been "shown in rigorous clinical testing to produce significant mind-body self-healing," a large minority of them will still get meaningful relief beyond what they get from usual care or warm, friendly appointments with doctors. Apparently, people are not averse to knowingly tricking their brains if it means there's a potential payoff. Some participants even insisted on getting another "prescription" for their microcrystalline cellulose pills when the studies were over. Colloca and other placebo researchers, including Benedetti, think this may open the door to an ethically acceptable use of placebos in medicine, in which doctors might prescribe a round of painkillers, telling the patient it contains both drugs and placebos but not revealing which pills are which.

When I asked Colloca why her outlook on alternative medicine is so different from Benedetti's, she told me it had to do with divergent goals of medical research. "When I worked at NIH, if I wanted to receive a grant, I had to think not just about the specific underlying brain mechanism, but the 'So what? How does this help patients?' We had to answer that question.

"In Italy, the system is quite different. You don't have to explain the relevance for improving the health of the country. You receive a grant just for designing a cool experiment. Fabrizio is interested in neurophysiological mechanisms."

Then Colloca mentioned something else that changed her outlook on alternative medicine. For many years while she was in Italy, she suffered from a seasonal allergic reaction to *Parietaria* plants, otherwise known as sticky weed. Her lungs would feel as if they were closing up and she struggled to breathe. She'd start wheezing and coughing and would rapidly try to suck in more air. Her doctors put her on a heavy dose of inhaled steroids, which helped her breathe but made her sluggish and didn't prevent her from being much shorter on breath than a twentysomething should theoretically be. She was unable, for instance, to make it up the stairs to her third-floor office without stopping.

But one day, a colleague who was an oncologist and an acupuncturist (in Europe it's not unheard of for MDs to study and practice acupuncture) offered a suggestion: "Let me needle you. Go off your steroids for a week and then come and see me. I'm sure you can feel better."

Colloca was appalled. She had never done acupuncture and didn't like needles, but far more worrying was her fear of going off steroids. Still, eventually she agreed to stop the drugs for three days and then get acupuncture. After the first session, which began with a needle between her eyebrows, Colloca was surprised to find that she could breathe almost normally. *Well, that won't last*, she thought, reasoning that it was probably a fluke. But it did last. None of her asthmalike symptoms returned. She never went back on the steroids. She did several more acupuncture sessions, and the next year when allergy season rolled around, she headed straight for the needles.

"Was this a placebo effect?" I ask.

"I think so, definitely. I hate needles, but I started to focus my attention not on my allergies but on this novel experience. I felt very relaxed, and maybe this feeling helped my autonomic nervous system release the constriction of my lungs and turn off my immune response. But we don't know a lot about the placebo and immune responses."

A few bits of data have emerged, though. One study found that a placebo pill significantly reduced allergic reactions in people with dust mite allergies. It's also been shown that people suffering from seasonal allergies, or allergic rhinitis as it's known, are sensitive to stress. Studies have exposed mildly asthmatic people to various allergens during times of stress, such as final exam week, and observed heightened immune or inflammatory reactions. Colloca says she was working as a postdoc in Benedetti's lab at the time of her foray into acupuncture, and that her workload and stress level were typical for an eager young scientist trying to prove herself—which is to say, fairly high.

It's less clear, though, whether the flip side of stress—relaxation and suggestions of relief, for example—can turn off immune or inflammatory reactions. The handful of studies that have been done on acupuncture and allergic rhinitis seem to show that the benefits are short-lived, not enduring like Colloca's. The American Academy of Otolaryngology (ear, nose, and throat doctors) does, however, recommend acupuncture as an option for allergy patients who want "non-pharmacologic therapy."

When Colloca told Benedetti about her miracle cure with acupuncture, she remembers him saying, "You're a very good placebo responder, Luana."

When I ask him about this, Benedetti is circumspect, saying that a single case is difficult to interpret. "It could be due to spontaneous remission or a psychological effect or the effect of an unidentified co-intervention or a specific effect of acupuncture or a bias, et cetera, et cetera. Simply put, we don't know."

Again, he's right. It's not possible to go back and figure out what exactly happened to Colloca's allergies, just as we can't say for sure why George O'Maille's postshingles pain went away. But I can't help thinking about how the need for scientific proof from controlled laboratory experiments and replicable double-blind, placebo-controlled trials sometimes butts heads with the real-life drama of chronic disease, where medicine unfolds a single person at a time and the right treatment can be a mystery revealed only through trial and error, if ever. When he got sick with multiple sclerosis, Dr. George Jelinek, an Australian ER doctor, realized that most patients are different from most doctors. "Studies which have not 'proven' the treatment to be beneficial but which suggest a major benefit look much more interesting when you actually have the disease, especially when the treatment has other benefits as well," he wrote in his book *Overcoming Multiple Sclerosis*. Dr. Jelinek chronicled how people can mitigate, though not necessarily cure, their MS symptoms through good nutrition, regular exercise, adequate vitamin D, and stress reduction.

He realized that when your body sidelines you from your life, you don't necessarily care what control group was used or what has or hasn't been published in the *New England Journal of Medicine.* You don't try to dissect the scientific plausibility of why a treatment is purported to work. When you're sick and nothing seems to be helping, you just want to find something, anything, that might release you from suffering's grip. And if it's possible for the tone and tenor of your interactions with another person to move molecules in your brain in ways that change the severity of your symptoms, then why not seek out confident, empathetic practitioners, even if they're dancing with proverbial chairs over their heads?

This is what Adam Engle, a man I met in what was then my hometown of Boulder, Colorado, did. After his health took a turn for the worse, he found himself in the company of a couple of self-styled healers—quacks, Benedetti might say. Adam wasn't sure if they could help him, but given his other options, he didn't think twice about letting them try.

Healing Partners

A case of vanishing back pain

The first time Adam Engle spoke with Peter Churchill, Adam remembers him saying, "I'll be able to tell you within one or two sessions whether I can help you or not." Adam found such a claim exceptionally curious. He'd been going to chiropractors and massage therapists for almost two years, and they hadn't helped him in any permanent way. His back was still a mess. Adam thought to himself, *How the hell can someone tell within one or two sessions if they can help me? How could he possibly know?* In addition to the chiropractors and massage therapists, Adam had also been to four different sets of doctors and surgeons, all of whom had told him in no uncertain terms that he needed surgery: the permanent fusion of at least two vertebrae in his spine and decompression to remove areas of bone that might be pressing against his nerves. Yet not only was this Peter Churchill not a doctor, he hadn't even gone to college. He was a largely self-educated therapist—a "healer" is how he was described—that Adam had come to through word of mouth. Several people had related remarkable results that they or someone they knew had achieved while working with Peter. Either some physical problem had improved or some emotional problem had dramatically shifted. One person called him "magical."

"I decided I had to at least see this guy, and I booked an appointment for the first free slot on his schedule," Adam says, telling me the story of his back problems in the sun-soaked living room of his home.

His pain had started, as it often does, with a sharp and inexplicable wave in an otherwise unmarred region. No particular moment ushered in its arrival; the problem was just mysteriously there, in the summer of 2010—a bolt of agony in his lower back and hips, and in his right butt a pulsation of tightness that seemed to radiate down his leg. A sixty-eight-year-old entrepreneur, Adam had endured backaches before—some 84 percent of us do—but nothing like this. Still, it wasn't crippling. Adam found ways to go about his life as usual and figured the problem would eventually go away, perhaps with the help of a few massages.

But when it was still there a few months later, Adam grew concerned. Trim and lean, he had always enjoyed exercise, going for weekly hikes in the mountains arising from his backyard in Longmont, Colorado, next door to Boulder. But now, rounding the corner toward seventy, he couldn't deny the realities of aging. Maybe this was how it was going to be, coming down the homestretch of one's life. Perhaps everyone reached a point when luck simply ran out and the years caught up with you. There would be less movement, less resilience, more aches and pains. After a productive life, now you were frail.

An MRI six months later, in December of 2010, appeared to confirm such suspicions. Adam's spine showed "moderate to marked" degeneration of the rubbery disks of cartilage that sat between the vertebrae of his lower back, and "moderate" spinal stenosis—a narrowing of the bony spinal canal, which houses the spinal cord. Adam's physiatrist—a physical rehabilitation doctor—thought this constriction could be putting pressure on a nerve and causing Adam's pain. The physiatrist mentioned physical therapy, epidural steroid shots, and surgery as options, but didn't seem to favor one over the others.

Adam knew he didn't want surgery. It seemed a drastic solution for a pain that wasn't horrible. He decided he would keep getting massages and start doing Pilates, hoping that strengthening his core

muscles might take some pressure off his spine. He also wondered if
this was just something he was going to have to live with.

Then, six months later, on a work trip to India, things got worse.
Moderate pain in his lower back and the right side of his hip suddenly
became shrill and piercing. Again, Adam hadn't been doing anything
unusual or inadvisably strenuous to trigger it. When he got home,
he added physical therapy to his regimen and found a chiropractor. A
Buddhist since his thirties, Adam also tried to manage his pain with
meditation, attempting to separate the irritation in his body from his
mind's attachment to it. But he managed only modest levels of fleeting relief. The following year, in 2012, two episodes of pain were so
shocking that Adam went to the emergency room.

"I was in a shit ton of pain and literally unable to transition
between sitting and standing," he says. "I couldn't get up from a chair
unless I went through a very contorted movement. It was like the
nerve in my right leg was getting hooked on something, so the pressure was just too great when I tried to stand up. Here, I'll show you."

Adam sets down a large bowl of almond-milk chai latte and walks
over to a chair. He is wearing a blue fleece vest over a white T-shirt
and looks like a man who might, at any moment, embark on a long
hike in the woods. I watch him hoist his butt off the chair a few inches
and bend forward at a ninety-degree angle. While folded up like that,
he swoops his head and torso around to the right as if scanning the
floor for an object.

"That would kind of unhook it," he says, once upright, "though
sometimes I couldn't do it without getting on the ground and rolling
around."

Adam sits back down on the sprawling beige sectional. In front of
us, a gas fireplace flickers, and I notice a bookcase dotted with Buddha
statues and a photo of his two teenage sons standing on either side of
the grinning, maroon-robed Dalai Lama. For two decades, Adam ran
the Mind & Life Institute, an organization he cofounded to promote
the scientific study of Buddhist practices. Before that, Adam, who has
a law degree and an MBA, had his own financial investment firm.

"I still didn't like the idea of surgery, but after the pain got worse I talked to some friends who said they had positive results from that, so I decided to get some surgical consultations."

The first surgeon Adam saw, in May 2013, delivered grim news. New MRIs taken at the Swedish Medical Center in Seattle, where his girlfriend was living at the time, revealed that his "moderate" stenosis had become "severe" in two places along his lumbar spine. Some bone spurs were also now on his joints, which, along with the previously noted disk degeneration, had caused two of his vertebrae to slip out of alignment in what's known as spondylolisthesis.

Each of the spine's thirty-three bones is grouped into a section—cervical, thoracic, lumbar, or sacral—and then numbered. The neurosurgeon's recommendation, which I read in a pile of medical records Adam had neatly assembled into a stack of folders on the couch, was to use screws and pieces of bone taken from Adam's hip to fuse together and stabilize his L4 and L5 vertebrae, the bottom two in his lumbar section, and to do decompression surgery to create a widening of his spinal column.

As dismaying as this advice was, it wasn't surprising. Both spinal fusion and decompression surgery are being used at historically high rates for people with back problems, especially in situations of apparent nerve impingement. Federal figures show that the number of spinal fusions in the United States rose from 56,000 in 1994 to 465,000 in 2011. Running at least $35,000 a pop, the procedure went from costing Medicare $343 million in 1997 to $2.24 billion in 2008, for an inflation-adjusted increase of nearly 400 percent.

The second opinion Adam got, at the Mayo Clinic, outside Minneapolis, was similar. But instead of fusing just L4 and L5, the Mayo neurosurgeons wanted to go all the way up to L2, where they also saw problems. "The surgeons were all the same," Adam tells me. "It was a twenty-minute consultation and the only difference was how many vertebrae they wanted to fuse. None of them ever touched me. They just looked at the images." He knew all these doctors had heaps of experience, but he was skeptical. Why were they so sure surgery

was his only option? A final surgeon he saw, in Boulder, also recommended fusion of L4 and L5, but when Adam pressed the surgeon, he issued a referral to a physiatrist at the Colorado Center for
Spine Medicine, who was willing to give epidural steroid injections a
try. This, the physiatrist said, might calm down the inflammation in
Adam's spinal joints and thus reduce symptoms.

When two injections had no effect, Adam asked the physiatrist
about other nonsurgical options. "He told me, 'You've got structural
problems and the only way this gets better is through surgery,'"
Adam recalls.

At this point, most people would have concurred. Three separate
teams of doctors had given clear recommendations for standard surgical treatments. But Adam had decided that he wanted to try a more
natural approach first. For more than two years, he'd been doing a
halfhearted, ad hoc rotation of massage, physical therapy, Pilates, and
chiropractic manipulations. This would be different. He was going to
fully immerse himself in a self-healing endeavor. He would stop traveling and taking on new commitments and focus on finding the best
people to help him, which he was fortunate enough to have the dis-
posable time and income to do, having just retired from his position
running the Dalai Lama–linked Buddhist organization Mind & Life.
"I figured I could give a more natural approach a try, and if it didn't
work, I could always go back and do the surgery. It wasn't like I had
a brain tumor and it was going to get worse and kill me."

Adam started weekly, hour-long sessions with Peter Churchill in
July 2013. Adam would lie on a massage table while Peter tried to
release some of the tension in Adam's muscles and tendons, sometimes through gentle massage and other times by thrusting his hands
into Adam's back and legs or digging into them with simple wooden
tools Peter had devised for this. Conversation often accompanied it;
Adam would tell Peter about his life, and Peter would ask what Adam
was feeling in his body in response to different things Peter was
doing. In some quiet moments Peter would close his eyes, take deep
breaths, and rest his hands on some part of Adam's body. On multiple

occasions, Peter asked Adam to imagine to himself becoming healthy again. Peter told Adam to pick a specific time, maybe a decade or two earlier, when he'd danced the tango and hiked mountains without any pain or worry. Remember what the experience was like and imagine yourself as that person, Peter said.

"Peter totally changed my way of thinking about this," Adam says. "He told me I could get better, but that I had to work at it. He views himself as a healing partner, where it's a journey together but you have to play your part and be open to it."

This, Adam realized, was what Peter was talking about when he'd said he could figure out quickly whether he could help Adam. It wasn't enough for Adam to say that he believed he could get better. He needed to feel this possibility deep in his bones and to think of his healing as a path he would follow wherever it might lead. "That was different from all those other people I'd seen. They were more in the paradigm of 'I'll do my thing and it will either work or it won't work,'" Adam says. A deep sense of trust quickly developed. Adam came to see Peter as possessing an impressive knowledge of the mechanics of the human body and a deep reservoir of intuition.

After one of the long quiet moments, Adam says Peter asked him to take a deep breath and then let it out. As Adam was exhaling, Peter pushed forcefully on Adam's lower back, eliciting an audible pop.

"He was able to reseat my sacrum," Adam explains, referring to the large, triangular bone at the base of the spine.

He believes that Peter—who has studied more than a dozen different alternative therapies, including cranial sacral therapy (which addresses the bones of the head), Rolfing (a type of deep tissue massage), and trigger point therapy (which focuses on tight areas within muscles)—was able to move other things, too. "He would gently vibrate over different areas, and it almost felt like he was teasing out the congestion in the muscles of my butt and leg and back, inducing blood and oxygen into the tissue, and removing scar tissue, which doctors will tell you is permanent."

Adam also found his way to Norm Allard, a chiropractor also

trained in a whole potpourri of practices—the Feldenkrais method, qigong, tai chi, and meditation, as well as forms of exercise such as Pilates, yoga, and martial arts. From all this, Norm had fashioned his own variety of exercise therapy for treating musculoskeletal pain. Based on gentle, small muscle movements and body awareness, his approach focuses on getting people to imagine and then feel connections in their body, such as among the tiny muscles linking all the vertebrae, or the links between the joints of the knee and the muscles going up and down the leg. The idea is to teach people how to get strong "from the inside out" and to foster a conversation with bodies they've been at war with. Norm does this in classes and also one-on-one with patients while they're lying on a table with limbs suspended in the air by a series of straps hanging from an overhead frame.

Adam gets up to show me some of the exercises Norm taught him. We walk across his living room to a small exercise studio he's outfitted with thick blue mats and inflatable balls. On one wall is a floor-to-ceiling mirror, and on the other, a scaffolding of wooden bars that looks like a ladder up to the ceiling for a person with short feet. Adam lies down on one of the mats, wraps his hands around a wooden rung, and tucks his knees to his chest. Then he rocks his legs from side to side, each time touching the floor in what he explains is a pelvic tilt.

I naively mention this looks easy. "Not if you have back pain. My partner's brother was here a few weeks ago and he couldn't even do it. He has back problems and his spine was so locked up. Plus, the way Norm does it, you focus on the movement of your head, whether your head is maybe rolling or skidding on the mat. You're getting exercise up and down the whole spine and connecting the whole thing."

Adam then sends his arms straight up in the air and swings them side to side in an upper spinal twist. He follows that with a rolled-up blanket tucked under his lower back and straightened legs catapulting down to the floor and back up again. This maneuver instantly takes me back to high school track practice, when we had to do such calisthenics and my legs shook miserably. Adam's legs, though, are perfectly calm and controlled. He actually seems to be enjoying himself.

That he can do such a thing so gracefully is a result of changes that occurred in the five months between July and November 2013. At first, the intensity of pain and the tightness in his lower back and the piriformis muscle in his butt eased, which allowed him to move more easily and without the need to pretzel himself up from a chair or tumble out of his car. Then the pain got even quieter and he started feeling generally stronger. By late November 2013—five months after he started one session or more a week with Peter and Norm, and seven months after surgeons and physiatrists told him he needed surgery—Adam realized that the pain, tightness, and numbness that had plagued him for three years was, in his estimation, 85 percent gone. All that remained, he says, was a good deal of tension in his right quad.

Now able-bodied and nearly pain-free, Adam was eager to start traveling again and had booked himself on an African safari. The prospect of long plane rides and jostling Jeep excursions was concerning, so he went to a physiatrist, a new one, to ask if he might need a back brace. The new MRI scans and X-rays she ordered showed, remarkably, that his spine was still a deteriorating wreck. "Degenerative spondylolisthesis with large facet effusions" and "right paracentral disk herniation compressing the L4 and L5 nerve roots," she wrote in her report. From the images, it looked like the spine of someone in a good deal of misery. But Adam no longer felt all the things that were wrong with his back.

The next time I speak with Adam, in August 2016, he has just returned from a month in Europe. The African safari happened with no back brace and without incident, and more recently he was sitting in a car for a twenty-five-hundred-mile excursion through Iceland and Norway, sleeping in a new bed every few nights and hiking nearly every day. "My back did not give me any grief," he says gleefully.

I ask Adam about placebo effects, whether he thinks his "healing partners" could have shifted his beliefs in a way that blocked or erased pain, and he hesitates. Nobody seems to like this word *placebo*, in part because it's never quite clear what anyone means by it. Is it a good thing or a bad thing? A tantalizingly powerful mode of self-healing, a middling and fleeting improvement, or an imaginary recovery no one should take seriously? I once had a conversation with a German acupuncture researcher that went on for twenty minutes before it became apparent that he thought I had meant the last, negative connotation instead of a more positive one. He went to great lengths giving arguments about why mind-body interventions have value, which I hadn't posed as a question.

This confusion, I think, reflects that the word *placebo* refers to something that isn't there rather than something that is, which makes defining it feel like trying to read a map printed in disappearing ink. By definition, a placebo has no specific or direct biological effects. Only what scientists refer to, in a great punting of the issue, as *nonspecific effects*, which is a term used both for lack of a better one and because it's difficult to track the neurobiological activity of such things as expectations, empathy, attention, and ritual with any precision.

It's also an open question whether the term *placebo* should be applied to any form of healing arising from a changed mind-set, such as the inner tranquillity of meditation or the acupuncture-induced relaxation and stress reduction Luana Colloca suggested might have contributed to her relief from seasonal allergies. Does a placebo effect inherently involve something hidden or misrepresented, or can we know when we're trying to use our minds to help us heal? Mindfulness expert Jon Kabat-Zinn once observed, "One way to look at meditation is as a kind of intrapsychic technology that's been developed over thousands of years by traditions that know a lot about the mind-body connection. To call what happens 'the placebo effect' is just to give a name to something we don't understand." Then again, maybe the term should apply more narrowly to observable expectation- and suggestion-driven symptom relief that can be reproduced in a lab, as Fabrizio Benedetti tends to argue.

Amid such linguistic confusion, a number of new terms have been proposed. The University of Michigan medical anthropologist Daniel Moerman has suggested the *meaning response*, and Harvard's Ted Kaptchuk nominated *contextual healing*. The retired Group Health researcher Dan Cherkin told me he likes *context effects*, and Paul Enck at Germany's University Hospital Tübingen said he thinks about *expectation effects*. There's also *empathy response, remembered wellness, belief effects*, and, confusingly, *MAC effects*, which stands for "meaning, context, and learning response." None of these seem poised to displace *placebo effect* anytime soon. As impoverished as the term is, it seems we're stuck with it.

Adam doesn't deny that placebo effects, or whatever you want to call them, played a role in his recovery, but he thinks Peter's and Norm's abilities go beyond the psychological to include the physical—the release of muscle tension, the elimination of scar tissue, and the reseating of his sacrum. He says his problems may have started with tension due to years of sitting cross-legged in a half-lotus position. But it's hard to see how this could be the whole story. Why wouldn't the bevy of massage therapists, physical therapists, and chiropractors Adam saw earlier have been able to address this tension or misalignment? Plus, the majority of the structural dysfunction in Adam's spine was still there after his pain disappeared, suggesting that we might need to look for answers in the mind and the brain as well as in the body. It's also possible that Adam could have experienced spontaneous healing that would have happened regardless of anything he did. I don't think this is likely, but such a coincidence is always a theoretical possibility.

I find Peter Churchill's Luminosity Healing Center at the edge of a dead-end street on the rural outskirts of town. Peter's large white stucco house emerges from behind an untamed thicket of trees, underneath which there is a trampoline and a neglected soccer net. It's a roomy parcel of land that opens up in the back to a full-fledged farm/petting zoo, with chickens, ducks, goats, and two horses quietly standing inside a barn.

The door I enter through is just off the garage and leads into a small waiting area filled with the steady sound of chirping birds, which I track to a noise machine stashed under a chair. I take a seat, not the bird-chirping one, and notice a carefully handwritten note placed on the table next to me—a plea from Peter's fifteen-year-old daughter, Serafina, for donations to help rescue and rehabilitate abused horses. This explains the animals I saw out back. I dig into my bag for a $20 bill to give to this seemingly sweet and enterprising child.

But have I really just formed this impression from the note? Before making an appointment to see Peter, I sat down to read his thick self-published book called *The Way Lightning Splits the Sky*. In it, he asserts his daughter possesses a "creative energy that can't be contained," a child who, as a seven-year-old, ran out of the house to put herself between her cockapoo puppy and two coyotes in hot pursuit of it. Standing defiantly in the yard, she let out a "mighty scream" that stopped the coyotes in their tracks and sent them off running, tails stuffed between their legs. I learned many other things about Peter's unusual life in this book, one that many of his patients read either before or after they start paying a nonreimbursable $150 per session to see him. The book plus word of mouth is the only way to know anything about Peter Churchill, who maintains a busy, full-time schedule. The totality of his digital presence is a static web page describing the book.

Peter's tale begins in the forests of suburban New Jersey, back when there were such things. He spent many childhood hours exploring and sitting quietly among the trees, waiting for deer, rabbits, and other creatures to appear. "I taught myself to still my thoughts by controlling my breath until my mind was free and uncluttered," he writes. "When I did this successfully, I found that the animals in the vicinity would suddenly begin to reappear from the fabric of the woods as if by magic, often walking right near me as if I were not visible."

After high school, Peter was accepted to nearby Rutgers University, but a visit to campus left him with an intense "feeling of confine-ment," and he decided to defer for a year and spend time skiing in the White Mountains of New Hampshire, hiking through the Great Smoky Mountains of North Carolina, and hitchhiking through Mon-tana. When one year was up, no way was he going to spend another four in a college classroom. Against his mother's pleas and amid threats of disownment from his grandfather, he decamped for a thin-walled summer cabin on a remote island in northern Maine, where he lived for nearly a year, hunting and gathering much of his food and testing himself, Thoreau-style, against the elements of a Maine winter. "I wanted to see what would happen through embrac-ing the intentional practice of suffering and solitude far from the comfort of civilization. I wanted to see what was underneath my usual positive state, to face my loneliness and sorrows and fears and see what might be revealed through the shadows on the other side," he writes.

After surviving intense loneliness and a night outside lost in a whiteout blizzard, Peter spent the next few years hitchhiking around the country with no particular plan other than to let the universe guide him to wherever he was supposed to be. He tells of taking rides when he was offered them and walking long distances when he wasn't, staying in someone's house if he was invited, and sleeping outside in a patch of woods if no invitation appeared, eating a meal if someone was sharing and fasting if people weren't.

Eventually Peter realized that what he needed was not a steady income and a roof over his head, but a spiritual teacher who could help him better understand life's big questions. This being the seven-ties, he found a ready supply of gurus and yogis and eventually made his way to Master Adi Da, whose real name was Franklin Jones. Franklin was raised on Long Island, but when Peter met him, he had become a self-anointed spiritual leader with a thriving community in San Francisco. Peter lived with Master Da and his devotees for over a decade before deciding it was finally time for that steady job and a regular place to sleep at night. Penniless and in his midthirties, he moved to Boston and started thinking about a skill he'd had since childhood—to pinpoint, he says, areas of pain and tension in other people's bodies and relieve them with his hands. Peter convinced two chiropractors to let him borrow their windowless back room, and he put out the word that he was doing "hands-on healing." At night, he read medical books and studied anatomy.

In fairly short order, he had enough clients to get himself an apartment with an extra bedroom in which to do treatments and then eventually a proper office. In the early nineties, he treated two patients of an internal medicine doctor at Harvard Medical School, Dr. David Eisenberg, an early adopter of alternative medicine. One of them had a five-year history of neck pain, and the other had suffered with sciatica for a decade. After seeing Peter, both returned to Dr. Eisenberg within a few days of each other with news that their symptoms were virtually gone. Intrigued, Dr. Eisenberg reached out to Peter and was sufficiently impressed to ask him to come and teach "hands-on healing for chronic pain" to Harvard Medical School students as part of an alternative- and complementary-medicine course, and also to present at an annual three-day integrative-medicine symposium hosted by Harvard and attended by doctors from around the country. Peter did this every year for nearly a decade. The move to Boulder came in 2007, when Peter and the woman he had married decided they wanted to live and raise children in a more mountainous and less urban setting.

The door to the treatment room opens, and a smiling middle-aged woman walks out, followed by Peter, in bare feet and with a ruggedly tan face. His long blond hair is slicked back and gives him the appearance of someone who has just emerged from a shower. His eyes flicker with activity.

Like Donna Eden, he has little use for formality. He heads straight in for a hug, then leads me into the treatment room, which is adorned with a busy array of spiritual, naturalistic, and medical effects. A large outstretched Buddha reclines across a long rectangular windowsill, its ample belly covered in bright green philodendron leaves.

On an antique wooden dresser are a colorful assortment of crystals and small brown bottles of what I assume are herbal remedies. Next to it, on the floor, are several plastic replicas of a human skull and a metal frame holding up a human spine.

I've told Peter I want to talk about Adam and also get a session for myself since I'm still not clear what precisely Peter does. But I'm trying to figure out what he should treat. I've been lucky to weather forty-nine years without any chronic health problems. The only thing that bothers me is a largely undiagnosed issue that happens from time to time when I eat. "Tell me about it," Peter says, sitting next to me and peering into my eyes.

I tell him it's a feeling that my throat is tightening or constricting. I'm not choking and it doesn't feel as if something is lodged in there. It's as though my esophagus has opted not to continue passing food and beverages through. Often, if I pause for a minute or two and take some deep breaths, it goes away. But every once in a while, about once or twice a year, it escalates and my throat feels as though it were completely sealed up. I can breathe, but swallowing becomes nearly impossible, which provides a real education in just how much saliva the human mouth produces. I have to walk around outside to keep myself calm before this terrifying experience finally goes away.

A number of years ago, a gastroenterologist stuck a camera down my throat to look for any abnormalities or, God forbid, tumors. There weren't any, and he suspected silent acid reflux, a condition in which you don't feel any of the typical heartburn symptoms and don't even know you have a problem. He wasn't certain though. He told me to take the reflux medicine Zantac to see if it helped. I never did because I was breast-feeding my youngest son and the drug passes through to breast milk. Even though the dose isn't considered dangerous, I wasn't in the mood to take chances and just filed away the uncertain diagnosis and decided to live with what was an intermittent and not debilitating condition. I don't imagine Peter can treat such a thing, but I'm willing to let him try.

As I talk, he sits still and listens. Then he has several questions:

Have I ever had the experience of choking on anything? Are there any specific food triggers? Is there anything that seems to help? I tell him yes, but long before this started; no triggers; and, outside of attempting to relax, no. "The throat is an area very susceptible to trauma because the body needs to protect the air pathways," he says. "Sometimes you can work on areas of the body to release that trauma." When I ask what trauma he might be referring to, he says it could be my first few experiences of feeling as if I couldn't swallow.

"So, my body has a kind of PTSD where past memories set off a panic?"

"Yes. It could be a physical pattern that sets up a more vigilant awareness, and the next time something feels a little tight, those cells react and pretty soon you have a pattern that seems unstoppable."

It's all very affirming and reminds me of the time I mentioned the problem to my new GP. She appeared thoroughly bored by the whole thing and suggested I could make another appointment to talk about it, presumably because the clock had run out on our brief time together. I left feeling as if my problem were a silly thing unworthy of professional attention. In just our few minutes together, Peter has managed to make me feel the opposite. He tells me to hop onto the massage table facedown. "I'm going to work the muscles along the spine, to get them to open up and help the central nervous system to communicate better." He presses one of the tools he used on Adam into my shoulder blade. It's basically a wooden dowel with a handle attached. "Any tenderness here?"

"A little."

He proceeds with the deepest deep tissue massage I've ever gotten. Then he tells me to flip over.

"I'm going to try to open these tight places, the throat and neck. Take a few deep breaths." With his hands, he starts to knead my shoulders and the sides of my neck. I bring up Adam.

"When he came in, I told him, 'Right now you're still convinced that you have these things wrong with your back, and you're still feeling like there's no cure except surgery. We have to completely change your mind-set around this or we have no chance.' I just felt that intuitively."

I'm now sitting up and Peter is dragging his fingers firmly along the length of my neck. Outside, a rooster crows.

"For me it's all about a relationship, the healing partnership. I think it's everything, the shared trust between two or more living beings. It doesn't mean you have to believe in what I'm doing, but you have to be open and want to have a shift toward being empowered and being more responsible and proactive in your own future well-being. It's a journey, and if someone doesn't want to be helped, I often can't help them."

Sometimes, he says, people can be coaxed into openness, such as the evangelical Christian who came in one day from Colorado Springs and did not seem to appreciate being glared at by a lounging Buddha. "She didn't say anything, but I could tell she wasn't comfortable," Peter tells me. So he gave her a handshake instead of the hug and tried to find common ground. He asked her if she had kids (she did) and if a love of the outdoors was what had brought her to Colorado (it had). Then he remarked that they weren't as different as she thought. "I tried to connect with her on some of the deep things we really did share."

I mention what Adam said about "reseating the sacrum" and the pop he heard. "I had been doing careful work to undo the lock that was happening between L4 and L5, the last lumbar vertebra," Peter tells me. "His sacrum was stuck in a position that was allowing it to dip into the rest of the spine, and I had gently been trying to move that back to where I knew it wanted to be, but it wouldn't move at all until he accessed the emotion that was stuck there. And when that happened, there was an audible pop that was extremely relieving to him."

This emotion stuck in Adam's spine, Peter says, involved a "huge personal betrayal" that impacted Adam's "whole identity as a man." He tells me I should ask Adam about it.

Later, when I do, Adam describes events that occurred at the Dalai Lama's compound in India in 2011, on that trip where his pain suddenly got worse. He had gone to Dharamsala for what was to be his last official event with the Mind & Life Institute and was surprised to be greeted by cold shoulders from some of the other participants and, most distressingly, the Dalai Lama himself. "It was a disastrous experience," Adam recalls. "I really wasn't made to feel a part of things." He later traced the chilly reception to rumors about sexual harassment and misappropriation of funds that a coworker had been spreading about him. The ordeal left Adam feeling unmoored and depressed. He talked at length to Peter about this.

I lie back down and Peter rests his hands along my head, moving them every few minutes. We stop talking.

"Okay," he says finally, in a way that signals my thirty minutes or so on the massage table has come to an end. "Take a deep breath and see what you notice."

I move my head around and swallow a few times. "It seems more open." But even as I'm declaring this, I'm also wondering if my throat has really opened up or if I've just fallen under the spell of suggestion again.

"It is. We created some actual room in there by opening all the different muscles around the throat. At the end, did you feel any warmth or gentle feeling as I held your head?"

I didn't, and in not doing so I apparently missed whatever healing energy he was sending from his hands. Peter waves this off. "It doesn't matter as long as there's a link of energy between two beings used with an intent to heal. The body uses that. It's a primal human sharing."

Peter also gives me some practical tips. I should pay attention to how I sit at my computer while I work, so that I'm not contorting my neck into a knot of tension, which he thinks is contributing to my throat problems. "I think you'll notice a difference. It might not be all gone, but it will be better."

Before leaving, I ask Peter if there are any health problems he doesn't treat. He quickly tells me there is one. "Cancer. It's so complex. When I was in Boston, people from Dana-Farber would refer patients to me, and I would try various things to see if it would have an effect on the actual cancer, and I found that it did not. I could help people with improving energy levels, pain, immune function, and a sense of well-being. But as far as the cancer itself, I don't even try anymore." Peter worked for several years with Leonard Zakim, the civil rights leader for whom the Dana-Farber Cancer Institute's Center for Integrative Therapies and Healthy Living is named. Zakim died from his multiple myeloma in 1999.

As I pull away from the house, it occurs to me that if one was trying to design an embodiment of the placebo effect—the positive one through which you can get real symptom improvement—you'd be hard-pressed to do better than Peter. Much like Donna Eden, he appears to have the whole package of everything placebo researchers say makes for those effective "therapeutic encounters"—the meaningful attention, expression of empathy, listening skills, earnest eye contact, steady projection of confidence, and an office adorned with a suggestive assembly of symbols. Add to that the Jesus-length hair, tales of a youthful quest for spiritual enlightenment, and what the pain researcher Richard Gracely at the University of North Carolina calls the "chameleon quality of a good clinician—the ability to see the world as a patient sees it, to understand the patient from within his or her frame of reference, and to communicate within this model." Moreover, I can't think of a better summation of how placebo effects might occur in the real world than "I can't help people if they aren't open to it."

Peter doesn't like the term *placebo* any more than Adam does. Peter thinks it reduces something profound and transcendent into a narrow, scientifically comfortable box. Sure, there were important changes in Adam's beliefs and the formation of a trusting partnership, but in his view it's more than this. "This is one of those healings where we shout out to the rooftops with gratitude and awe. There was a miraculous component to the whole process that did not spring from any one of the various parts, but from something much greater and indefinable—the Sacred," Peter writes to me later in an email, capitalizing the *S*. There were also, he says, "higher energetic healing frequencies" transmitted from him to Adam, which "changed the cellular environment of his injured tissues," along with the "high art" of skillful manual manipulation of Adam's cranium, sacrum, and lumbar spine.

I'm not sure what to think of "healing frequencies," but I do know that Peter's hands-on manipulations resemble the kind of spinal adjustments you get at a chiropractor's office, the deep tissue massages given by massage therapists, and the joint and muscle handling done by physical therapists, all of which fall into the broad category known as manual therapy. The unifying idea is that doing specialized manual maneuvers to the body can help sort out issues in the musculoskeletal system—joint immobility, bone misalignment, tension in the soft tissues—in ways that ease pain. This would seem to be a more reasonable and probable factor in Adam's recovery than those "healing frequencies." Recently, scientists have been making headway on this question, the potential of manual therapy for chronic pain, but they're finding out that the story isn't quite as straightforward as it seems.

My Back Is Out

The hand therapy of chiropractic

The first chiropractic treatment took place in the town of Davenport, Iowa, in 1895. It was administered by a heavily bearded fellow named Daniel David Palmer, a Canadian who had moved to the Midwest with his family when he was twenty. Palmer had been a schoolteacher, grocer, and beekeeper before deciding upon a career as a "magnetic healer," an eighteenth- and nineteenth-century form of energy healing that, confusingly, didn't have all that much to do with magnets. Created by the German doctor Franz Mesmer, "animal magnetism" held that the difference between living and nonliving things was an invisible animating force, a magnetism, that could have physical effects including healing. In treating patients, Palmer would pour his animal magnetism into someone, sometimes without touching them. The whole idea was famously debunked in a Paris experiment ordered up by King Louis XVI and overseen by Benjamin Franklin, among others, in which a blindfolded young volunteer tried in vain to find the apricot tree that one of Mesmer's disciples had magnetized. The boy hugged all the wrong trees and, before he could find the right one, collapsed in a fit of magnetic rapture.

After nearly a decade of magnetic healing, Palmer yearned to find his own method of working with patients—hopes he realized one day after talking to his office building's janitor, Harvey Lillard, about his hearing problems. According to Palmer, Lillard told him that seventeen years earlier, while bent over in a cramped and unbalanced position, he'd felt something give way in his back, producing an instant loss of hearing. Palmer described Lillard as "deaf," but he also said the two men had a conversation, so it's unclear what sort of hearing problems were at issue. Palmer was intrigued by Lillard's story and began searching for clues, quickly finding one. "An examination showed a vertebra racked from its normal position," he wrote in 1910. "I reasoned that if that vertebra was replaced, the man's hearing should be restored."* Given the nineteenth century's understanding of medicine, such a conclusion was not unreasonable. The briefly popular "reflex theory" held that nerves running in and out of the spinal cord were a control center for all bodily organs, including the brain (as opposed to the other way around). An impinged spinal nerve could be messing up one's ears, heart, lungs, liver, and any number of other things. Palmer concluded that Lillard's misaligned vertebra was blocking the nerve controlling his ear, thus impeding the flow of both nerve signals and the body's *innate intelligence*—Palmer's updated term for animal magnetism.

He set to work trying to get Lillard's vertebra back into place, pressing his hands firmly into several areas of Lillard's thoracic (upper) spine until Palmer felt the offending bone pop back into alignment. "I racked it into position by using the spinous process [the bony protrusion at the back of each vertebra] as a lever and soon the man could hear as before," Palmer wrote.

Excited by these results, he did more "spine racking" on other patients, including one with unspecified "heart trouble." He wrote, "I examined the spine and found a displaced vertebra pressing against the nerves which innervate the heart. I adjusted the vertebra and

---

*In a letter written eight months after this encounter, vastly sooner than Palmer's remembrance fifteen years later, Harvey Lillard said he had never considered a link between his back and his hearing until Palmer made the association. "Dr. Palmer told me that my deafness came from an injury in my spine. This was new to me; but it is a fact that my back was injured at the time I went deaf," he recalled.

gave immediate relief." Palmer, who did not elaborate on the symptoms the patient was getting relief from, believed he was onto something. "I began to reason if two diseases, so dissimilar as deafness and
heart trouble, came from impingement, a pressure on nerves, were
not other disease due to a similar cause?"

Had Palmer known that the nerves responsible for hearing are
tucked safely inside the skull and do not run along the spine or that
the vagus nerve controlling the heart travels down the neck and into
the body without touching the spine, he might not have jumped to
such lofty conclusions. But such knowledge about the nervous system was not yet widely recognized.

To Palmer, ideas about spinal misalignment and nerve impingement looked like the answers to healing he'd been searching for. Like
the ancients who blamed demons whenever sickness appeared, Palmer
suspected that a single factor was responsible for all illnesses and that
when it was found, the wretched curse of bodily disease would finally
be lifted from the human race. In the "done by hand" therapy of *chiropractic*, a term one of Palmer's friends coined from the Greek words
for hand (*chiro*) and practice (*praktikos*), Palmer imagined an approach
"destined to be the grandest and greatest of this or any age." He began
teaching others how to manipulate and adjust spines to cure virtually
any affliction, not just ill-defined hearing problems or heart trouble.
He claimed that "95 percent of all diseases" could be traced back to
the way misaligned or "subluxed" vertebrae pressed on nerves, causing them to tighten, increase their "vibration," and impede the flow of
innate intelligence. Palmer started teaching his methods, and by 1902
fifteen chiropractors had graduated from Palmer College of Chiropractic, which is still humming along today in Davenport. It also has
campuses in San Jose, California, and Port Orange, Florida.

Palmer wasn't the only starry-eyed medical entrepreneur on the
American frontier. In Kirksville, Missouri, a day's drive from Davenport, a doctor had grown disgusted with his drug- and surgery-
obsessed colleagues. "My vote is now, first, last, and all the time,
against the use of anything but Nature's remedies for treating the

sick," Dr. Andrew Taylor Still wrote in 1910, during an era in which the standards for becoming a physician were far less rigorous than they are today. Still watched in horror as well-intentioned colleagues took patients through all kinds of ritual abuses—bleeding, purging, burning, and blistering—and gave them medicines containing arsenic, mercury, opium, and castor oil. He also heard tales of what he considered to be the unnecessary removal of body parts performed in unsanitary surgeries, a "growing curse," he declared, that was "mutilating the body and throwing away that which is useful and should be retained as a part of the human body for its longevity and comfort."

Still argued for a holistic focus on the patient rather than the disease and began a method of manual joint manipulation he called osteopathy (*osteon* being Greek for "bone"). It was like Palmer's soon-to-follow chiropractic, but instead of impinged nerves causing disease, Still believed misaligned bones were obstructing arteries. In 1892, he set up the American School of Osteopathy (now A. T. Still University in Kirksville) to train students in what he called the "very sacred science" of restoring blood flow by realigning bones.

Some have speculated that since Palmer emerged several decades after Still, Palmer must have derived some of his ideas from Still. But any such debate obscures that neither man invented joint manipulation. The practice appears throughout history and stretches back to the ancient Greeks. Hippocrates devised elaborate tables with straps, wheels, and axles to move patients' spines around, and Galen was known to stand or walk on dysfunctional backs. In the Middle Ages, European "bonesetters" passed manual healing skills down through family generations. After pronouncing a joint to be "out," they would seek to snap it back into place, with an audible crack. In other areas of the globe, the lomilomi massage practitioners in Hawaii and the shamans of Central Asia had their own styles of vigorous manipulations, both of spines and areas of soft tissue throughout the body.

Today osteopaths are doctors—DOs, or doctors of osteopathy—who go to separate, lesser-known medical schools, where, in addition to taking the same courses on anatomy and physiology that MDs do, they get instruction in manual methods of manipulating joints, muscles, and connective tissues, or fasciae. Although sometimes considered alternative, osteopathy and its ninety-six thousand US practitioners have become relatively accepted members of the medical community, thanks to the profession's willingness to adapt to the advance of medical knowledge in the twentieth century. Osteopathic physicians provide the full range of services that MDs do—writing prescriptions, performing surgery, doing imaging, etc. (This isn't true, however, in Europe, where DOs aren't physicians and focus primarily on manual manipulation.)

Chiropractors, on the other hand, stayed truer to their late nineteenth-century roots and thus incurred the ire of the medical establishment. In the first few decades of the twentieth century, hundreds of them were sued for practicing medicine without a license, including Palmer himself, who went to jail for seventeen days before deciding to cough up the fine. Chiropractors in California adopted the battle cry "Go to jail for chiropractic," and in 1915, at the height of the controversy, 450 of them did. Then in the sixties, with chiropractors growing in numbers, the powerful American Medical Association tried to put a stop to the whole business. Calling chiropractors an "unscientific cult," it urged its members not to associate with them or refer patients to them. Palmer's devotees, who had been busy getting licensing laws enacted in all fifty states, did not take kindly to such aspersions. Chiropractors filed a lawsuit accusing the AMA and several affiliated medical groups of trying to "destroy a licensed profession." In 1987, a federal judge ruled in their favor, declaring that the medical groups had engaged in a conspiracy of "systematic, long-term wrongdoing." Instead of getting stamped out, the profession continued to grow. In 1990, the US had forty thousand chiropractors; today, there are seventy-seven thousand.

I'd never been to a chiropractor before and I decided I should try to find one the way everyone does, by asking friends and consulting random strangers on Yelp. Since I'd recently moved to Honolulu due to my husband's job change, the friend ranks were a little thin, so I leaned more heavily on virtual ones, who had nothing but glowing things to say about chiropractor Dr. Gary Bell. (Some chiropractors call themselves doctors because they receive a "doctorate of chiropractic" in chiropractic school, but this isn't the same thing as being a physician.) Dr. Bell's website includes a lengthy bio that describes him as an avid surfer, a devoted dad of two adult kids, and a professional who believes in commonsense things such as moderation and preventive medicine. He is not among the minority of chiropractors who still subscribe to Palmer's idea that an innate intelligence or some other vital life force inhabits the body. Like most modern practitioners, he also does not treat "95 percent of all diseases," but instead a narrow band of back, neck, and occasionally other musculoskeletal complaints. One Yelp reviewer describes his "kind eyes," and another declares, "Seriously, the best doctor visit I've ever had."

I book an appointment when the neck soreness that visits me from time to time has gotten particular noisy. Rotating my head from side to side, I can hear a snapping and crunching as if sheets of ice are giving way underneath my skin. This tension and achiness comes and goes, usually in accordance with the amount of time I've spent hunched miserably over my laptop, and it's not what I would call chronic neck pain. Most of the time it doesn't bother me, but it's exactly the sort of thing that, along with back pain, chiropractors are supposed to be able to address. Online, I've seen videos of them wielding numerous unusual devices and performing fancy-sounding techniques to alleviate conditions like mine, and I'm curious if this more modern version of chiropractic will work for me. Possibly Palmer was onto something even if he was wrong about all the ways his methods were supposed to work. It isn't preposterous, after all, to imagine that moving around spinal joints and surrounding muscles and tendons could help with neck and back problems.

Dr. Bell's office, on an increasingly bustling commercial strip in Honolulu, has the bland, generic feel of a dentist's office—white drop ceilings, universally appealing artwork, and obligatory stacks of magazines. When he ushers me into one of the treatment rooms, he takes a few minutes to examine me intently, watching me turn my head from side to side and poking his fingers around in my shoulder. Then in his calm manner (and with what I suppose are kind eyes), he declares that my neck is quite straight, which he assures me is not a desirable trait. Neck bones are supposed to stack up in a slightly curved fashion so they can better distribute the round pile of heaviness set on top of them. "Our heads weigh ten to eleven pounds," he says. Dr. Bell thinks my problem is the result of trouble in my C4, C5, and C6 vertebrae. "Vertebrae are supposed to move, but yours are stuck. Everything is too tight." Again he pokes his fingers around the contours of my neck.

To alleviate this, Dr. Bell first attaches strips of electrodes to the underside of my neck. This sends warm and prickly jolts of electricity to my skin, which is supposed to loosen up my muscles and tendons. I can't seem to decide if this feels good or annoying, and I conclude that somehow it's both. After my neck has been sufficiently electrocuted, he moves on to a standard method of spinal manipulation. While lying on my stomach, I take deep breaths, and as I'm exhaling, Dr. Bell heaves his palms into half a dozen spots on my back, each time producing a gratifying crack or pop. A few areas, though, require multiple attempts: "Ooh, you're tough. Lots of tension. You have an athletic body type and your body wants activity, but you have long periods of inactivity when you're sitting." He urges me to get up periodically from my hunched position and start exercising more than a meager once or twice a week.

On my second visit, Dr. Bell employs some of those devices I've seen touted online. First, there's an impulse activator, which resembles a gun. He places it along my back and neck and pulls the trigger, eliciting a series of loud, sharp pops. *Baap, ba ba bap. Baap, ba ba ba baap.* I am surprised to find these gentle thumps of pressure satisfying. "It overstimulates the area and you get a good reflex that actually helps the area relax and the muscles loosen up," he explains. I also lie down faceup on a "spinalator" table that sends three giant rollers swooping languidly along my back. Then I go facedown on a moving spinal decompression table, which rhythmically pivots the lower half of my body toward the floor. This moving table, Dr. Bell notes, is his favorite device, one designed to "open up the disks."

As a final treatment, he tries to get my displaced C4 through C6 neck vertebrae back into alignment by turning my head to one side on the table and then pushing on spots on my neck while I exhale. Despite multiple shoves, nothing is moving. No pops or cracks are forthcoming, and he decides not to keep trying, for which I am grateful since I've read that neck manipulations carry a small risk of complications. On occasion, they can make someone's pain worse or, more tragically, tear a vertebral or carotid artery and cause a blood clot that travels to the brain, causing a stroke. Such occurrences, although real, are thankfully rare.

After both of my nearly hour-long sessions with Dr. Bell, I notice that my neck feels smoother, lighter, and less sore. For the remainder of the day, it's like having a new neck. But the next morning the creakiness settles back in, and I am forced to conclude that chiropractic didn't work for me, at least not as I would have hoped. Studies have shown, however, that spinal manipulations of the sort Dr. Bell did can be modestly beneficial for some people's necks. And a compilation of all the higher-quality research seems to indicate that having someone manipulate your spinal joints is about as good for chronic back pain as doing exercise therapy, taking pain meds, or getting a massage (which entails the pushing and kneading of muscles and other soft tissue rather than the manipulation of joints). Both the American College of Physicians and the American Pain Society recommend spinal manipulation as a treatment to consider for moderate and short-term benefits when low-back pain does not improve on its own. "Manipulation is as good as conventional medical care for back pain and probably safer given the side effects of drugs and injections and surgeries that we do," Dr. Richard Deyo, a professor at Oregon Health and Science University and an expert on back pain, tells me.

The intriguing question is why. Echoing Palmer's theories, many chiropractors say they are correcting "vertebral subluxation," which causes the jamming of nerve communications between the body and the brain by a misaligned back, neck, or hip joint. But no scientific backing exists for the idea of clogged nerve signals. Maybe it's the more reasonable and evident explanation that chiropractors such as Dr. Bell, along with many osteopaths and physical therapists, give. How moving joints back into alignment with specialized movements—much as Peter said he did with Adam—can relieve undue pressure on surrounding areas and thus ease pain. Or how the softening of specific muscles and tendons can coax more blood into afflicted areas. Both of these mechanical explanations seem like rational descriptions of how these treatments might be helpful for chronic back and neck pain. But are they?

As I'm looking into this, I come across Joel Bialosky's story. Now a clinical associate professor at the University of Florida's physical therapy department, Bialosky practiced as a physical therapist for years before realizing that something wasn't quite right with what he had learned. When I speak with him over the phone one afternoon, he tells me he became interested in physical therapy during his senior year of high school because, as he put it, he liked sports and excelled in science. He had no idea what a physical therapist did. "A bit of a blind squirrel finding the nut. I got lucky and it turned into a career I have really enjoyed."

Although they weren't apparent as such at the time, Bialosky's first clues came while he was in grad school at the University of Pittsburgh. In both his graduate and undergraduate classes, he had been taught that for any given problem you had to find just the right technique to address it. There were the basics of soft tissue massage for the muscles and tendons, and for the spine, the different "velocities" and "amplitudes" of manipulation determined how quickly your hand pushed and the depth, as well as the direction and distance, it traveled. Every technique had numerous subtechniques. Not to mention all the different devices and spine-plying tables like those I enjoyed in Dr. Bell's office.

Yet whenever Bialosky watched his mentor at the University of Pittsburgh work in the clinic, he always saw him use the same one or two methods for the vast majority of patients, regardless of their problem. Bialosky wondered how so few approaches could be effective for so many different issues. Then at the physical therapy centers Bialosky worked at after graduate school, more things didn't add up. He would often fail to agree with colleagues on which vertebrae were out of alignment on a given patient or how best to address someone's knee inflammation. Everyone seemed to be making slightly different assessments of what was wrong and what the right treatment was. "I was very focused on identifying and stretching tight structures and strengthening weak ones," Bialosky says. "But it got to the point where when someone got better, I never knew whether to give myself credit or not. I never knew what was really making a difference."

Curious, he started digging into the scientific literature on manual therapy, which encompasses both the manipulation of joints and the massaging of soft tissue. In the few studies that have compared one type of spinal manipulation or massage to another—thrust versus non-thrust, high velocity versus low velocity, advanced therapeutic massage versus relaxing massage—there was little difference in the outcomes. Much as real acupuncture doesn't outperform superficial sham needling, each form of manipulation seemed to produce similar levels of pain relief and improvements in mobility. The intricate specificity of it all didn't seem to matter. Also, when a practitioner was allowed to evaluate a patient with back pain and then carefully choose a personalized treatment, the pain-relieving effect was no greater than when the practitioner performed a randomly assigned treatment.

Bialosky now began to understand why his teacher had clung to those few tried-and-true methods. All things being equal, he had picked the techniques that felt most comfortable to him and that his patients seemed to like. With a good deal of dismay, Bialosky realized there wasn't much point to all the elaborate maneuvers he had spent years learning and perfecting. "You can still go spend all kinds of money to learn all kinds of techniques that don't seem to matter," he says ruefully.

There were more disheartening revelations. According to studies, manual therapists are not good at using their hands to objectively assess the position of different vertebrae. When blinded to a patient's condition and one another's conclusions, they don't come to much agreement about which vertebra is supposedly out of alignment on a given patient. In one study done at the Arthritis Research Center in Wichita, Kansas, therapists were asked to figure out whether patients had fibromyalgia, a puzzling condition in which everything hurts and nothing seems to be wrong, or myofascial pain syndrome, a chronic pain that causes sensitive points in muscles, or whether they had both or neither. In evaluating how the therapists did, the researchers wrote, "It was a disaster. The examiners were distraught."

"What I discovered," Bialosky says, "is that there's not much evidence for the whole biomechanical model. Chiropractors and other therapists will often say they're going to try to do specific techniques, directly to a vertebra, to put it back into place, but that's not what's happening." Although joints and bones do move around when you push on them, he says these structures go back into their original position shortly thereafter. That popping noise therapists work so hard to summon—the ones my back emitted in Dr. Bell's office, that came with Peter's "reseating" of Adam's sacrum, and that was so critical in Palmer's treatment of his deaf janitor—is not the sound of joints sliding back into their rightful position. We know this because several years ago physical therapy researchers at the University of Alberta used MRI machines to peer inside cracking joints. They found that when bones on either side of a joint are forced apart, a tight bubble of superslippery joint-lubricating fluid forms and then quickly releases, spraying tiny bits of fluid into the joint cavity. "Backs do not go 'out' and then back into place," Greg Kawchuk, the lead researcher, wrote in an email. "It's an old myth perpetuated by many professions as they think it's an easy way for patients to understand why they have pain. It's a fairy tale." Kawchuk, who was a chiropractor before going into research, did his experiment on knuckles, but he sees no reason why the same concept wouldn't apply to all joints. He's not sure why exactly this sudden spritz of lubricating fluid feels so relieving, but speculates it's because of the way it stimulates the tiny nerves inside a joint.

In Bialosky's view, using manual therapy for chronic musculoskeletal pain—regardless of what muscles or joints are being pushed or pulled or rubbed or in what manner—is less about correcting some particular area of the body and more about changing the mind and the brain. "It's all those contextual factors," he says, referring to the complex set of beliefs, expectations, feelings, and assessments that arise from a treatment—the story we construct around an experience: Does that oh-so-satisfying pop make you feel that something is getting fixed? Do you believe your chiropractor always knows the exact spot you need to have adjusted? "If you came to see me for an intervention, and then you went with the exact same problem to another person who gave you the same treatment, there's a potential we might produce a dramatically different outcome based on how you and I interact and how we get along."

Bialosky sometimes refers to this common underlying mechanism as a placebo effect, though it includes something that's rarely mentioned in discussions of placebos: our brain's response to the physical sensations created by someone's hands (or impulse activators, strips of electrodes, spinalator tables, and the like).

"Our view is that the mechanical force from manual therapy initiates a cascade of neurophysiological responses from the peripheral and central nervous system," Bialosky says, speaking of nerve impulses that travel from whatever area a therapist is manipulating or massaging, then up through the spinal cord and into the brain. "The current thinking is that it's not really the specificity of that force, but how those signals tone down the sensitivity of the nervous system and lead to a reduction in pain."

I ask how a spinal twist or hamstring massage can "tone down the nervous system."

"Your guess is as good as mine," he says, explaining that research into this is not robust because up until recently there has been little reason to study it. For the most part, nobody has questioned the work of physical therapists or chiropractors such as Dr. Bell. Doctors happily refer chronic-pain patients to them and insurance routinely covers it.

To better understand what might be happening to the body when it finds relief in manual therapy, it is illustrative to circle back to acupuncture, another therapy that produces a steady stream of bodily sensations—in this case from needles and sometimes an acupuncturist's hands—but which has been the subject of far more research. Unlike manual therapy, acupuncture's mechanisms have never gotten a free ride. Meridians and qi enjoy a high public profile because scientific explanations for acupuncture have always felt as if someone is picking science terms out of a hat—the "stimulation of endorphin release," the "activation of A-delta nerve fibers." At the Cleveland Clinic, Dr. Neides told me about "manipulating positive and negative ions in a way that can reduce inflammation." I've also seen meridians discussed as "tendinomuscular" structures and regions of "increased temperature and low skin resistance." For the past two decades, a semi-organized group of acupuncture-curious scientists from around the world, many at high-impact organizations, have been trying to find data-backed, pseudoscience-free explanations of what sticking needles into people really does and why it can relieve pain. Some of them even have an intelligible hunch about the biological mechanism.

This Feeling in My Body

How acupuncture really works

Inside a cabinet in the lobby of the Martinos Center for Biomedical Imaging is a porcupine's nest of thin, bright green circuit boards. They protrude upward from a domed mold of plastic, their bits of silver shimmering in the beams of light from overhead. It's a surprisingly alluring glimpse into the gear that surrounds one's head when lying inside one of science's most popular and potent tools for peering into the human brain. First developed in the early nineties with the help of Martinos Center scientists, fMRI machines (the *f* stands for "functional") track blood flow in the brain with the help of huge magnets and radio waves. When these waves bounce off oxygen molecules in the blood, they spit out signals revealing where the brain is the most active at a given time, with oxygenated blood a presumed measure of the brain's exertion. Although future generations of scientists will undoubtedly regard such devices as crude forays into what neuroscientist David Eagleman has called "three pounds of the most complex material we've discovered in the universe," today's fMRIs are the best shot we have at understanding the mysteries of the human brain, with its 80 billion neurons and its several hundred trillion–plus possible connections between them.

Located along the Boston waterfront in the historic Navy Yard, the Martinos Center was established in 1989 after the daughter of a

Greek shipping magnate committed suicide at MIT. Grief stricken and hungry for answers about the mental illness from which his daughter suffered, Thanassis Martinos donated $20 million to Harvard Medical School and MIT to start the center. Since then, scientists here have made breakthroughs in understanding mental health, created new treatments for epilepsy, researched opiate addiction, and, much to the chagrin of skeptics, used fMRIs for acupuncture research. Over the past twenty years, Martinos Center scientists have gotten some $25 million in research grants for acupuncture research, much of it from the National Center for Complementary and Integrative Health (NCCIH) unit of the National Institutes of Health. In blog posts in 2013 and 2015, Steven Novella, a neurologist at the Yale School of Medicine and vocal skeptic of alternative medicine, declared this to be a complete waste of money, the equivalent of dumping bags of cash, tealike, into Boston Harbor.

"The very concept of acupuncture adds nothing to our understanding of the universe, and biology and medicine specifically." Then he added, "In my opinion humanity should not waste another penny, another moment, another patient—any further resources on this dead end. We should consider this a lesson learned, cut our losses, and move on."

From behind his desk on the second floor, Vitaly Napadow seems only mildly annoyed when I bring up such criticisms: "Acupuncture has a lot of baggage." He sighs. "It's almost like it suffers from its two-thousand-year history. Because of meridians, because of qi, these things that are prescientific, these pseudoskeptic people out there think that if acupuncture is accepted in medicine and covered by insurance, then it means that we're also saying that meridians and qi are real. If something is effective, there could be all kinds of different mechanisms, so why not be open to it? I think acupuncture helps people, and I'm interested in finding out how it does that."

Napadow is a slight, mild-mannered fortysomething with a goatee and an earring. He is an assistant professor at Harvard Medical School and runs the Center for Integrative Pain Neuroimaging at the Martinos Center. He is also a licensed acupuncturist. Every Thursday at noon, he leaves the Navy Yard to drive west into the suburbs of Chestnut Hill, where he treats four or five patients at the Brigham and Women's Pain Management Center. "It's fibromyalgia, back pain, migraines," he says. "We get lots of patients who have failed all the first-line stuff—the drugs, the injections, the surgeries. Then they come to get acupuncture."

A Ukrainian immigrant who moved to Baltimore with his parents when he was six years old, Napadow's introduction to traditional Chinese medicine began with tai chi classes while he attended Cornell University. He liked the way the synced breathing and slow, mindful movements made him introspective and quietly conscious in a way he had never before experienced. "I've tried to meditate, but I'm not very good at it," he says. "It's hard for me to just sit there." An interest in acupuncture didn't come until he was at MIT getting a master's degree in mechanical engineering. He was also then attending Harvard Medical School as part of a joint medical-engineering program with MIT. Immersed in the intricacies of human biology and physiology, Napadow found himself wondering about Chinese methods of healing, with which he was only passingly familiar. This curiosity didn't arise from any particular experience with acupuncture. "I literally went to the New England School of Acupuncture, the oldest school of acupuncture in the US. That's here in Boston. I told them, 'I don't know anything about Chinese medical theory. What is yin and yang? What is qi?' I asked if I could audit a class."

They told him he couldn't, but that he could do the program half-time. "So, during the day I was doing all this coding and studying structure-function relationships in muscles, and then at night, I was learning acupuncture."

Napadow didn't find this as jarring as you would think. It was as if medical science and acupuncture lived on two different planets inside his brain. "Some things from Chinese medicine are just wrong," he concedes. "But the appeal is that Chinese medicine is really such a holistic, thirty-thousand-foot view of the body and how it works.

I think the way modern medicine and modern medical research is structured, we're very much in the trees and we miss connections between the different systems. I also loved the patient interactions and the way it feels like the art of medicine. The poetry and the philosophy of it all." As for such things as meridians and qi, he thinks of them metaphorically, as "philosophical entities."

After getting his PhD in medical engineering and master's in acupuncture, Napadow wasn't sure what to do next. He could leave research behind to work full-time as an acupuncturist, or he could continue doing basic biomechanics research and forget about acupuncture. But Boston is one of the places in the US where medical research and alternative medicine coexist comfortably. In 2004, Harvard Medical School offered him a faculty position in radiology, he started treating patients at the Brigham and Women's Pain Management Center, and he continued as a postdoc at the Martinos Center, where he was part of an ambitious research project to find out why an ancient, prescientific practice could sometimes help people, especially with pain.

The Martinos Center's first inquiries into this began in the late nineties and predated Napadow's arrival by a few years. To see how the human brain looked on acupuncture, researchers wheeled people into the tunneled chambers of fMRI scanners and put needles into specific acupuncture points. At first, work done here and elsewhere that seemed to offer scientific validation for the uniqueness of acupuncture points created great excitement. A study done at the University of California, Irvine, and published in 1998 in the *Proceedings of the National Academy of Sciences*, for instance, showed that when acupuncturists needled the foot along the bladder meridian, the brain's visual cortex lit up. In Chinese medicine, these points are used to treat eye-related disorders. Maybe there was some hormonal or chemical pathway we didn't know about by which needling the foot could send messages to the brain and then into the eyes or other organs.

"There have literally been hundreds of these poke-and-look studies," Napadow says, glancing briefly at one of the three large monitors barricading his desk. "The first generation of them were limited because they didn't control for points on the body or for needle insertion, so you couldn't tell whether the location mattered or whether it was important to insert the needle." The researchers on that bladder meridian study eventually realized that there were many possible explanations for their results. Merely closing one's eyes can, paradoxically, activate the brain's visual and auditory cortexes. No fancy stimulation of the foot is necessary. In 2006, five of the eight researchers retracted the paper.

The other limitation of many of these poke-and-look brain imaging studies, Napadow says, is that they have been done on healthy volunteers who get a single session of acupuncture, which doesn't necessarily tell you anything about what the practice does for people with chronic health problems. Napadow's group and others are now embedding brain imaging into clinical trials of acupuncture, so that they can, it is hoped, connect the dots between brain changes and symptom improvements.

I follow Napadow around hallways and into and out of elevators while he shows me, with a kind of fatherly pride, the Martinos Center's six fMRI machines. Today, he was supposed to be scanning a woman who had spent six weeks getting acupuncture treatments as part of a fibromyalgia study, but she had to cancel. So instead we peer inside unoccupied electromagnetically shielded rooms as he boasts of their varying tesla levels. (*Tesla* as in the unit of magnetic field strength, not the car.) Visually, the machines look indistinguishable from the coffin-like MRIs hospitals use. The difference is that while MRIs give you a static picture, fMRIs are like a live satellite feed from your brain.*

As we traverse the halls, Napadow explains that the more recent and better-controlled studies that he and other acupuncture researchers have done with these machines have had more conclusive results.

---

*It isn't quite this simple; fMRI machines spit out reams of data that researchers then use software to crunch into a real-time picture of brain activity.

One is that both types of acupuncture used in clinical trials—the properly inserted needles that are twisted to produce a pulling or tingling sensation, and the superficial poking with fake, blunt-tipped needles (or toothpicks)—produce a similar imprint of brain activity. Specifically, they both activate brain areas important for the processing of tactile sensations, such as the somatosensory cortex and thalamus, and they deactivate brain regions known as the default mode network. You can think of this network as the daydreaming brain or the neurological basis of our sense of self. It's active when you are silently ruminating on your life or otherwise tuning inward. This default system deactivates when you're engrossed in accomplishing a task or focusing intently on something happening in the external world, when you "lose yourself in something."

That poking tiny needles into people gets them to register bodily sensations in ways that mute the mind's chatter and direct attention into the body shouldn't be surprising. But in Napadow's estimation, these similarities in brain activity indicate that pokes with retractable needles or toothpicks aren't inert placebo treatments after all, a conclusion now shared by most of his colleagues in acupuncture research. "We actually have way more sensory receptors on our skin than we do deep within the muscle and fascia, and when you poke someone, you're activating these receptors," he says. Put another way: both real and fake acupuncture needles are doing something neurological. But what?

"I think it has to do with context and meaning," Napadow says, singing the same tune I heard from the physical therapy researcher Joel Bialosky. "Now, if I brush up against a table as I walk past, it won't heal my back pain, but if these palpable sensations going from the skin receptors to the brain are coupled with the brain expecting something meaningful to happen during a complex healing ritual, then it's processed in very different areas in the brain versus some sensation that goes up to maybe those same brain regions, but is kind of ignored or only barely perceived at a conscious level."

Napadow is saying, in a roundabout way, that real and fake needling are both active ingredients in a type of sensation-enhanced placebo effect—where feeling things in your body is an important part of believing it in your mind. The needles aren't necessarily creating some specialized nerve signal or chemical reaction at the site where you're inserting them, at least in the case of pain. They are more like visceral alerts to the brain that healing is on the way, though how exactly such messages work to alleviate symptoms is a big unknown.

However, just to make things more confusing, Napadow also believes that real and sham acupuncture are working through slightly different pathways in the brain. In 2009, he was part of a study that used radioactive PET scans to look at neurotransmitter activity inside the brains of people with fibromyalgia. Led by Richard Harris at the University of Michigan, the study found that sham acupuncture looked a lot like the known placebo mechanism of endorphin release that Howard Fields and Jon Levine first discovered at UCSF in the 1970s, whereas real acupuncture looked like some other self-healing process in the brain—perhaps, the authors suggested, an increase in the number of receptors on the surface of neurons that endorphins and other neurotransmitters can attach to. Whether you could also call this a placebo effect was unclear.

Although I'll buy Napadow's theory of acupuncture as a "somatosensory-guided, mind-body therapy," it's still just that, a theory. He is currently completing a study on people with low-back pain that he hopes will further isolate the role needle sensations play, but he has yet to analyze the results. Also still a hypothesis is his view that where you place acupuncture needles matters "a little," probably due to higher densities of sensory receptors in certain areas of the body. "I think we can say it's not to the fine degree that's taught in the textbooks," he says.

Since we're speculating, I ask Napadow whether the concept of a bodily placebo effect for pain might apply to other types of therapies. Why, after all, would sensations from needles be so uniquely special? Could Bialosky's "neurophysiological responses" from hands have similar effects? And what about gentler forms of touch such as Reiki and Eden Energy Medicine?

Napadow says it's certainly possible there would be similar pathways for other kinds of touch, but their effectiveness will likely depend on the condition being treated. More intense and slightly painful sensations, such as those from acupuncture needles, spinal manipulations, and deep tissue massages, might be more useful for chronic musculoskeletal pain than light touches, which could be helpful for depression or anxiety. Napadow has shown in several studies that electro-acupuncture—sending low currents of electricity through needles for greater levels of stimulation—is effective for carpal tunnel syndrome. But because the mechanism here is not just the signals themselves but how our brains register them, it's likely that the narratives we weave around different therapies and the people who deliver them will play a big role.

Given the preliminary nature of all this, I am surprised to find that researchers have related theories about how exercise helps the brain change the way it responds to messages from the body. This is somewhat counterintuitive because we know for sure that exercise is specifically useful for our bodies—it strengthens muscles, increases the oxygen efficiency of our lungs, stabilizes blood sugar metabolism, and so on. But for easing existing chronic musculoskeletal pain (as opposed to preventing it from happening in the first place, which strengthening the body can clearly help with), none of this seems to matter. When one type of exercise is compared against another in studies of chronic pain, usually back pain, the most perennially studied, there doesn't seem to be much difference. Strengthening your trunk muscles is just as good for alleviating back problems as the McKenzie method of spine stretches, which is to say modestly so. Walking is just as helpful for pain from knee osteoarthritis as quadriceps strengthening. One reason for this is that researchers have yet to figure out how to reliably identify subgroups of patients who will, in fact, respond better to specific types of exercise. It is assumed that, for example, those with spinal instability might do best with Pilates or yoga, whereas those with deconditioned muscles should do walking, but no one has been able to prove this in large, high-quality studies. The other reason for the lack of specificity is that, much as with manual therapy, there appears to be a common, brain-based mechanism for many types of exercise.

"It's like your nervous system is learning how to override pain" is how Luciana Gazzi Macedo, an exercise researcher at McMaster University in Ontario, explains it. Instead of manual touch or needles cuing a healing message to the brain, Macedo says that exercise can change a person's frame of mind—for example, reducing stress and anxiety or generating positive appraisals about one's physical health or conditioning—thus rewiring how the brain's pain-processing areas handle bodily signals.

Such rewiring is most likely to happen, says Jim Rainville at Boston's New England Baptist Hospital's Spine Center, when exercise hurts a little—"tolerable discomfort"—and when people realize this doesn't mean they are damaging themselves. "In order to desensitize the neurons in the pain system," he tells me, "you have to give them repeated positive stimulation, and then the patient has to be indifferent to that stimulus. Mind-set is critical."

So Henry Beecher, the midcentury placebo researcher, was right. Pain wasn't entirely what it seemed, and there were more ways of addressing it than people assumed. But why is this symptom always the one alternative approaches seem to have the greatest effectiveness for? How can pain be so malleable to ginned up expectations of relief and the context in which you come to feel it or seek to relieve it? What is this vexing sensation, anyway?

8
. . . .

Brain Pain

The modern neuroscientific view of pain

Back when I was at the Cleveland Clinic, I sat in a circle with seven other people and a therapist. Everyone except me, and presumably the therapist, was in some form of physical pain—backs, knees, shoulders, headaches, and the all-over hurt of fibromyalgia. It was the same class George O'Maille had taken several years earlier with such unexpected results. We had just finished some chair yoga and a positive-message meditation ("May you know wholeness and happiness in your life just as you are" was the gist), and now Kellie Kirksey, one of the Center for Integrative & Lifestyle Medicine's ebullient counselors, was leading a discussion about how everyone had been doing that week. She was asking for examples of achievements, setbacks, and strategies that might have been useful.

Carol, an outgoing woman in her sixties with back, neck, and shoulder problems, led the way by saying that since the last class she had increased her dose of pain medications. This appeared to be movement in the wrong direction, but Carol went on, rather jarringly, to talk about what a great week she'd had. She said she'd been using the yogic breathing techniques to help her relax when her pain started getting loud. Over the phone, a colleague had told her she sounded "one hundred percent better."

"Normally I would just dismiss this," Carol said triumphantly.

"But instead I told her, 'Thank you.' It was the first time I didn't beat myself up for feeling good."

Candace, who wore a rumpled plaid shirt and looked to be in her late twenties, had more explicit milestones to report. For the first time since her fibromyalgia started three years ago, she had enjoyed several nights of eight uninterrupted hours of sleep. "My boyfriend said it looked like I didn't move all night," she said, beaming. She thought it had to do with breathing techniques and visualization exercises they'd learned, which she said helped her relax and reduce her anxiety. Her fibromyalgia pain, a seven out of ten when she'd started the classes, was now a four.

Most moving was Deborah's testimony. A trim and put-together woman in her forties, Deborah said she had fibromyalgia plus a history of migraines dating back to her childhood. She talked about how another migraine could strike at any moment and how its crippling pain could lead to multiple days of fevers and convulsions. Decades of such erratic and unpredictable suffering, she said, had left her with a kind of "trauma" that had become an illness itself, an affliction that kept her from living the life she wanted. Deborah also told us about her mother: Individual hypnotherapy sessions that were part of the classes revealed how her mother had encouraged Deborah to think of herself as a vulnerable victim in need of care. During such sessions, Deborah was put into a light trance and encouraged to delve into experiences that had caused emotional suffering. Deborah now believed that these disempowering views about herself were contributing to her migraines and fibromyalgia. "I know for a fact that deep-seated unconscious beliefs are a big part of this for me," she declared.

Chronic pain doesn't often get flagged as a serious medical condition the way stroke, heart disease, cancer, or a MRSA infection does. It doesn't kill us and it's hardly a unique experience. The only people who haven't ever felt pain's barbed edges are the tiny number of those born with what's called congenital insensitivity to pain, which sounds like a good thing but definitely isn't. People with this genetic mutation have no idea when they're injuring themselves. They're known to unwittingly rest their hands on hot surfaces, walk on broken legs, and suffer hearing loss from untreated ear infections. Often, they don't make it beyond age twenty-five.

Chronic pain is a major medical problem. The National Institutes of Health has determined that a staggering number of Americans—50 million of us, or one in five adults—suffer from significant or severe daily pain. This includes people with some underlying causal condition, such as cancer, multiple sclerosis, rheumatoid arthritis, ankylosing spondylitis, diabetic neuropathy, or heart disease, as well as those with chronic pain as their primary complaint, sometimes without an obvious cause. Back pain is the most prevalent of these, but there is no end to body parts that can hurt without any clear or fixable medical reason. Such problems are so prevalent they have their own names—irritable bowel syndrome (gut pain), tension headache and chronic daily headache (head pain), osteoarthritis (knee pain), temporomandibular disorder (jaw pain), interstitial cystitis (bladder pain), complex regional pain disorder (arm and leg pain), fibromyalgia (all-over pain), and vulvodynia (vaginal pain). All totaled, both the obvious and more ambiguous cases of chronic pain are equivalent to the combined number of people who suffer from cancer and diabetes.

After the chronic pain class, I chatted in the center's lobby with Connie, who was waiting to get an acupuncture session. A lively woman in her sixties, Connie had taken the pain classes in 2015 and, like Deborah, felt that hypnotherapy sessions had identified lifelong patterns Connie suspects have contributed to her health difficulties. "Growing up, there was no affection in our house, no physical contact, and you had to be perfect to get our parents' attention," she said. "So I think that's why when I found out about my husband's issues, I thought, 'Well, you don't deserve better than this because you're not perfect.'" By "issues," she was referring to her husband's many extramarital affairs, which she found out about many years later and two children into their marriage.

Connie looked fit and energetic and was wearing pink yoga pants, having just come to the center from a spin class at her gym. She said she didn't necessarily think her emotional stress had caused her to develop Cushing's syndrome, an adrenal gland disorder, but she was certain it had extended some of its symptoms over two decades. As she started to feel more psychologically grounded, her physical state seemed to improve. "In the pain class, we all started feeling successful. They built us up and made us believe in ourselves again. That class gave me my life back," she declared.

It was hard not to see the pattern. Patients were linking their emotional well-being to their physical pain—Connie and her philandering husband, Deborah and her codependent mother, Carol and her self-esteem. But were these things really connected?

There are indeed clear links between psychic pain and bodily discomfort. We know that fighting with our spouse can dial up a headache, for example, and that a looming deadline can turn a sore back into a throbbing mess. An increasing number of studies are showing that managing stress and negative thoughts can help reduce pain in a large minority of people. Cognitive behavioral therapy (CBT), a form of goal-oriented talk therapy that's becoming a standard non-pharmacological treatment for chronic pain, teaches patients how to identify the unhelpful thoughts and behaviors arising from pain—for example, "My pain will never stop," "I just can't take it anymore," or "I'm not going to exercise because it hurts too much"—and then turn them into positive ones, such as, "This pain flare-up is temporary." According to studies, CBT helps roughly 40 percent of people achieve a 30 to 40 percent reduction in their pain, at least for some amount of time. "Pain is unavoidable but suffering is optional," a cognitive behavioral therapist might say.

Mindfulness-based stress reduction uses meditation to help people find greater peace of mind. Acceptance and commitment therapy does something similar, though with more talking, less meditating. An even newer approach, called emotional awareness and expression therapy, deals head-on with difficult emotions. In a 2017 study, patients with fibromyalgia were encouraged to identify suppressed feelings through exercises in group therapy, including role-playing and creative writing. Those that did this for eight weeks had substantially better physical functioning and less widespread pain than patients who got fibromyalgia education classes. Those in this newer therapy saw similar overall improvements as those in the CBT group, but more of them got dramatic (greater than 50 percent) pain relief—23 percent of patients versus 8 percent for the CBT group.

Yet, given how the visceral sensation of pain manifests in our bodies, so insistently rooted in some area of our flesh and bones, this makes little intuitive sense. Why treat the mind if pain is in the body?

Lorimer Moseley seems like a stereotypical Australian. He hikes in the outback, tells *yarns*, and uses such words as *reckon* and *mate*. He also seems to be having more fun than most pain scientists. At conferences, his talks are peppered with amusing asides, such as the time he and a colleague ran down a hallway with a "borrowed" prosthetic arm from the orthopedic department. Or the time he got bit by a venomous snake in the outback and woke up, still in the wilderness, three days later. In one presentation, he refers to his children as Lord Adorable Squeakypants and Princess Petunia-Cake.

"Not their real names," he notes.

Before becoming a pain scientist, Moseley was a practicing physical therapist, or physiotherapist as they're called in Australia. He's now a professor at the University of South Australia and one of a growing number of researchers trying to unravel the complexities of the universal experience known as pain. Moseley catches my attention because he's asserted that *pain signals* don't exist. This common phrase is still used all over the place, in pain videos, tutorials on the Web, government websites, and scientific journals. Just the other day, I stumbled upon this tangled bit of locution in the respected *Nature Reviews Neuroscience*: "These [calcium ions] both enhance the excitability of spinal cord neurons in response to incoming pain signals and cause an exaggerated release of neurotransmitters from sensory neuron presynaptic terminals to the spinal cord."

"Well," Moseley says when I reach him on Skype, "it's a term of convenience, I think. And we've known this for some time. I'm writing letters to journals where the term is used saying it's mislabeling, and the person who wrote it would agree with me." Illuminated by a fluorescent ceiling light, Moseley's nearly bald head appears far larger on my screen than it must in real life. "My opinion is based on my appraisal of the current body of knowledge, and it's not simply a semantic debate. Talking about 'pain signals' keeps people from a real understanding of how pain works. I think it's really compelling that if you don't have a brain, you can't have any pain. Pain is the most powerful protective mechanism we have, a harm alarm. It won't hurt if your brain doesn't think you are in danger and in need of protection."

Moseley seems to be saying that pain is never inevitably dictated by some area of the body by "pain signals," but is a result of our brain's evaluation of what's going on in the body and what we should do about it. This is the case for any type of pain, he says, whether acute or chronic.

But is this true? My mind jumps to an incident a few weeks earlier when I slipped on a wet area in the garage and, instead of inelegantly toppling over, managed to steady myself with a great slam of my big toe on the ground. Pain immediately radiated from this recently pedicured body part, and under the chipped nail, a small, ominous-looking puddle of blood formed. "My toe was clearly relaying something up along my nerves, and it sure felt like pain," I say.

"The language we could use to describe that," Moseley says patiently, "is that there is a very dangerous or injuring stimulus in your foot, a sudden change in pressure and perhaps chemical balance, and this sends a barrage of signals to your brain telling it about all these new changes. But no card-carrying scientist would say that that barrage is 'pain.' These messages sent along the nerves to your spinal cord, or what we call nociception, only say 'danger,' not 'pain.'"

Still not getting it, I ask how my brain could have opted not to feel pain.

"Well, if someone fired a gun behind you at the same time you dropped something on your toe, you would not feel pain."

Moseley's probably right: A gunshot would be a far more menacing threat for my brain to grapple with, and I might not initially feel my mangled toe at all. You could call this a distraction, but it can also be understood as a relative danger evaluation.

"The point is, there's just not a one-to-one relationship between tissue damage and pain."

Intertwined with the immense complexity of our brains, pain can be fickle, slippery, and erroneous. This is known, he claims, from several lines of research. In an early nineties study exemplifying one such line, a team of psychologists at Baylor College of Medicine in Texas attached electrodes to the heads of sixty volunteers and told them they would be receiving electrical currents from a "shock generator." This, the psychologists said, "was safe but often painful." (If you're wondering who would volunteer for such an experiment, the answer is unemployed and underemployed men who were being compensated for their services.) The trick was that for some of these cash-strapped individuals the shock generator was never turned on. Yet, amazingly, half of this group experienced pain anyway. The higher the large visible dial on the fake machine was cranked up, the more severe these men rated their pain. Although their torment was real, it had no basis other than their brain's belief that their head was hooked up to a creepy torture device they thought was on. The brain perceived danger to the body, and so it created pain.

"We just have to convince people we are doing something dangerous and we can make it hurt," Moseley says.

Not long after this study was published, a bizarre real-world example of this kind of sourceless pain occurred in an emergency room in Leicester, England. In an article in the *British Medical Journal*, doctors describe a twenty-nine-year-old construction worker who jumped from a scaffold and landed directly on a large nail, which promptly disappeared into his boot. When he arrived at the ER, the man was in such terrible pain that even the smallest movement of the nail was excruciating. To remove the nail without causing further distress, the doctors sedated him with fentanyl and midazolam. Once the man was no longer screaming, the doctors removed the nail from his boot and pulled the boot from his foot. You can guess what they found: The nail had completely missed the target; it penetrated the man's shoe right between his toes. His foot had barely been scraped. Phantom limb pain, where people feel agony and spasms from amputated body parts, is a similar phenomenon, though one far more intractable than a fake foot stab since it comes from long-standing neural connections in the brain that don't disappear just because a limb does.

From other studies—and from Henry Beecher's wounded World War II soldiers—we know that the opposite scenario can also take place: There can be damage but no pain. In 1994, doctors at Hoag Memorial Hospital in California did a study that stunned nearly every spine doctor who heard about it. They took spine MRIs of ninety-eight men and women with no history of back pain and found that 64 percent of them had at least one herniated disk—the sort of adulteration thought to cause back problems but that obviously doesn't in many people. Other studies have revealed that even pain-free twenty-year-olds in the prime of their lives had what would seem to be bad backs. Thirty-seven percent had some form of disk degeneration. Yet, much like Adam Engle with his herniated disks and degenerative spondylolisthesis, these people were not bothered by their structural damage.

Similar studies have revealed that knee X-rays can also be unreliable predictors of pain. Between 20 and 60 percent of people with no knee problems have significant joint deterioration, occasionally even with cartilage so eroded that it's almost bone rubbing on bone. Jaw imaging too can be misleading. Although dentists have long believed the pain of temporomandibular disorder, or TMJ, to be caused by a misalignment of jawbones, a recent study found that about half of pain-free jaws look out of sorts on an MRI.

Conversely, between 10 and 15 percent of people with relatively normal knee X-rays are in pain. In some baffling cases the X-ray of the right knee looks like a disaster, but it's the left knee that hurts. Back pain experts now consider about 90 percent of all cases to be "nonspecific," meaning a clear anatomical cause can't be identified, which goes a long way toward explaining why spinal surgery is often ineffective for back problems and why vertebroplasty, in which bone cement is inserted into a fractured vertebra, has been shown to be no more effective for back pain than a placebo incision and a belief that the procedure has taken place.

Moseley does not mean to suggest that chronic pain doesn't involve bodily damage. Nearly always, he says, there will be some "issues in the tissues," whether inflammation, lactic acid buildup, or deconditioned tissues that are being slightly overloaded. But such peripheral problems, he maintains, often don't tell the whole story, especially in those cases where doctors aren't able to identify a clear cause.

A driving feature of much chronic pain, he says, is a nervous system that's become oversensitized, like a smoke alarm screaming at you every time you burn some toast. This dialing up happens in the cells of the spinal cord and the brain and causes our system for modulating bodily sensations to go off-kilter. Brain cells that send "on" messages into the spinal cord become more active, while those that signal "off" messages grow quieter, making those pathways for relaying danger messages up to the brain overly sensitive. "When this happens," Moseley says, "the brain is being told there is more danger in the tissues than there really is."

Over time, this barrage of misleading and magnified dispatches creates a rewiring of neurons in the brain's pain-processing areas, making pain easier to produce. "The longer you have pain, the better your system gets at producing it," he says, glancing down at his phone, which has beeped six different times while he's been talking, sending life's current version of a danger message. Yet like those people with brains unperturbed by degenerating spines, he ignores it and continues talking.

He tells me his group in Adelaide recently tested people with self-reported "back stiffness" and found that they didn't have any more measurable stiffness than people who were free of spinal complaints. What they had was a propensity to overestimate the forces being applied to their spine, an unconscious "perceptual error." Similar tendencies have been observed in patients with irritable bowel syndrome and fibromyalgia; the amount of pressure needed to make their guts or thumb hurt is far lower than it is for pain-free people.

"This science isn't boring, it's amazing stuff," he asserts. "Why on earth haven't we embraced this amazement more? That amazes me!"

I tell Moseley that such findings, rousing as they are, conjure up the bromide "It's all in your head." If pain is our brain's evaluation, right or wrong, of bodily trouble, then does this suggest we're bringing some of our chronic pain problems on ourselves? People with chronic pain often fear their doctor is harboring exactly this view. In her book *The Body in Pain*, the Harvard scholar of English Elaine Scarry writes that while our own pain is incontestably certain, accounts of other people's silent suffering reflexively engender doubt. No objective test yet exists to identify pain the way tests exist for infections, blocked arteries, damaged nerves, or cancer cells. Doctors pretty much have to take your word for it.

"Well, we don't use that phrase [*all in your head*]," Moseley says, "because the pain people feel is absolutely, one hundred percent real and legitimate, and we want to get people to understand, really understand, that in the belly of their nervous system. I would say instead that it's all *of* your head, and that what's happening is that your brain is engaging this truly amazing protective mechanism that's there for you, even if your tissues are okay. In times when your body really is in danger, it can pinpoint the exact spot that you need to stop and pay attention to. Pain is so compelling like that."

I marvel at Moseley's ability to turn a negative connotation into a selling point. But, given the wide range of problems that can be tagged as *chronic pain*, I suspect it's difficult to determine to what extent someone's pain is coming from a significant bodily problem or from a skewed brain assessment—the distress signal or the response to it.

For more information, I talk to Dr. Dan Clauw, the director of the Chronic Pain & Fatigue Research Center at the University of Michigan. Speaking on his cell phone in his car, he tells me about a questionnaire he uses to determine whether someone has what he calls "brain pain." This self-report survey asks patients about how widespread their pain is and to what extent they have other central nervous system problems such as fatigue, trouble sleeping, and issues with memory and mood. Clauw says that under this measure between 20 and 80 percent of chronic pain patients turn out to have brain pain, with conditions such as fibromyalgia, irritable bowel syndrome, low-back pain, TMJ, and the bladder discomfort of interstitial cystitis on the upper end of that percentage range. Cases with known damage or inflammation, such as arthritis and diabetic neuropathy, are on the lower end.

"We're not telling doctors not to do knee replacement surgery," Clauw says. "We're just saying, here are some simple self-report measures you can give to determine the extent to which someone's pain is coming from their knee or their brain. If it's the latter, then that person is five times less likely to respond to knee replacement surgery or opioids than another person who seems to have a better nervous system volume-control setting."

What drives all this brain pain—what makes someone's degenerated disk or dissolved knee cartilage hurt like hell while another person's stays quiet as a mouse—is still an enigma. But there are compelling theories. Vania Apkarian, a neuroscientist and pain researcher at Northwestern's Feinberg School of Medicine in Chicago, has done research showing that the brains of people whose low-back pain becomes chronic have a specific signature of changed neural activity as compared to those whose back pain gets better. In chronic sufferers, instead of nerve signals occurring primarily in the thalamus, which is the brain area responsible for relaying sensory input from the body, there is heightened activity in the centers involved in emotion, decision making, learning, and memory—or in neuroscientist-speak, the amygdala, medial prefrontal cortex, nucleus accumbens, and hippocampus.

I call Apkarian to ask him if this means that emotions, beliefs, and past experiences help determine whether someone's pain dissolves as a temporary inconvenience or calcifies into lasting misery. Because if they do, then Carol's self-esteem, Connie's guilt, Deborah's learned self-pity, and even the Dalai Lama's cold shoulder to Adam could be important factors in their illness.

"There is no other alternative [to our results]," Apkarian tells me. "If these limbic brain regions are the drivers of the reorganization of the brain, those are the areas that determine our psychological state and define our emotional personality. So in this sense, it is the baggage of our personality that we bring into our experience of the injury and its interpretation of that injury that creates or protects against becoming a chronic pain patient. You have this cascade of emotionally driven learning events that effectually reorganize the brain into a chronic pain state."

It's also probably the case, Apkarian says, that certain people have a genetic predisposition to react to pain in unhelpful ways due to genes that code for the activity of certain neurotransmitters in the brain, though not much is known about this widely held hypothesis. Either way, Apkarian says, whether some cases of chronic pain are linked to our baggage or our genes, they represent a brain "addicted to pain," a peculiar concept he knows raises more questions than answers. "This is a new topic and we're talking about billions of neurons, but we are moving forward. We have theories and a hypothesis that we didn't have even a few years ago. For years, the pain field was dominated by the simple idea that the nerve fibers and the spinal cord circuitry are really the main reason for pain, and that if you can control those, you will control pain, but that's just not the case."

Apkarian's work syncs up with observational studies that show one of the biggest predictors of persistent pain is someone's mental and emotional state when it begins. According to these studies, if you go in for surgery, suffer an injury, or just randomly start feeling agony somewhere, your pain is much more likely to continue if you are also feeling significantly depressed, anxious, bummed about your job, isolated from friends and family, sensing that you have no control over your life, or feeling feeble and freaked-out about what your pain means.

"There is a constant battle in the brain between whether you should respond to pain and sensations in the body or go about other activities," Howard Fields, the longtime UCSF pain researcher and Apkarian's coauthor on some of his studies, tells me. "The minute pain starts, an individual is going to have a set of unspoken expectations about what it means, how bad it is, and how long it's going to last. These have a lot to do with cultural and previous experiences of pain, with what you've been told, and with the state of your body and mind at the time."

Fields has his own parlance for chronic pain—a "learning mechanism"—though, like Moseley, he is careful not to suggest anyone should be blamed for their pain, any more than a heart disease patient should be chided for needing a stent operation or a multiple sclerosis patient be implicated in the destruction of their myelin sheaths. "Your brain is biologically based and you're definitely not to blame," Fields says. "I have this idea for a T-shirt. It says, 'It's not my fault. My neurons made me do it.'"

I must be a sucker for nerd humor, because I'd definitely buy that T-shirt. Fields's point is that, much like the formation of disease in our body, signal amplification in the brain and spinal cord happens below the surface of consciousness. Although we may be wise to some of the contributing negative beliefs or painful memories, we aren't usually aware of how these things could be rewiring our brains to perpetuate pain. Like so much of what our brains are up to, it happens without our knowledge or consent.

Although the pain field is still awash in unanswered questions, at least we have a general understanding about how the sort of emotional confessions I heard at the Cleveland Clinic might be therapeutic for someone's pain. Whether it's that, CBT, mindfulness-based stress reduction, acupuncture, or Peter Churchill's "healing partnership"—it's not a mind-over-matter scenario. None of these things hold any direct dominion over a knee joint, spinal disk, or shoulder tendon. Instead, they are capable of shifting our states of mind, such as our stress and anxiety levels and the stories we tell about ourselves about what our pain means. To treat chronic pain, sometimes you can go with surgery or drugs to fix the body and mute the barrage of messages being sent to the brain about damaged tissue. Other times, nothing is substantially wrong in the tissues and you need to rearrange matters in the brain. Increasingly, this is also being done with drugs that target brain neurotransmitters, such as antidepressants and antiepileptics, such as the gabapentin doctors prescribed for George O'Maille's postshingles pain.

When I talk again with Moseley, he says that trying to alter mindsets is a big part of what he does in the pain clinics he consults for around the world. "We have to keep reminding people in chronic pain that they have these huge buffer zones. You hurt but you're safe. So, we try to load every part of a patient interaction with safety cues." By this he means subtle messages coming from someone's tone of voice, facial expression, eye contact, and touch, anything that builds trust or expresses enthusiasm and empathy—the same things that can unleash placebo effects, though Moseley isn't going to break the trend of dislike for this term. "There's a presumption that placebo effects aren't biological or effective, when what we're capturing are the real effects of things we haven't measured or are not aware of yet. This is where alternative medicine is way ahead of medicine in my view, because over time we've downplayed these skills. We've had these drugs that are so biologically active, and people have misattributed the effect of everything they do to a pill. We're now starting to realize we should capture and exploit these effects."

Moseley realizes that feeling "safe" is an intensely personal and mysterious calculation; one person's safety cues are another's waste of time. The underlying neural circuitry trapping us in chronic pain doesn't yield easily to the forces of willpower, rational thought, or good intentions. If it did, people who experience phantom limb pain could get rid of it by gently reminding themselves that they don't have an arm to hurt, but this rational approach doesn't work. Some people in chronic pain try everything under the sun and still ache, and whether this is because they haven't tapped into the thing that's going to rewire their brain or because of some undiscovered or untreated bodily problem is rarely clear. "I've had plenty of scenarios where no matter what I do, I can't move someone," Moseley says. "But I've also seen people stick with something and get big changes after years of trouble. Or long after the fact, I run into patients whom I thought weren't making ground only to see them out and about and thrilled to tell me how well things are going. People really are fearfully and wonderfully complex." He pauses. "And then there's always miracles."

Miracles?

"I saw a miracle patient and I thought it was all me. She had a twenty-five-year history of back pain, and I did my usual thing for two hours and then I said, 'See you in a week.' A week later, she came back, out of her wheelchair, looking ten years younger, with a sprightly step, and I said, 'So, you're looking pretty good. Do you want to tell me about that?' I was waiting for her to sing my praises, but instead she told me about what she did after she saw me."

The patient, Janet, went to see what she described as "her sister's clairvoyant," with whom Janet had been eagerly waiting for an appointment for six months. Her reasons for going had nothing to do with her back, but as she was leaving the session, the clairvoyant stopped her, locked eyes, and said, "Janet, there's nothing wrong in your back anymore." Janet was greatly surprised. She hadn't mentioned her back the whole time, though surely a crystal ball wasn't required to correctly guess spinal issues in someone arriving in a wheelchair. When Janet woke up the next morning, her pain was completely gone.

"We followed her for thirteen months and it didn't come back," Moseley marvels.

He doesn't mean to argue for clairvoyants as an overlooked group of back healers. The tale illustrates how sometimes an unexpected message, delivered to a particular person in a particular state of openness, can be a kind of miracle. This particular "shock to the mind," as Galen would put it, wouldn't have worked for 99 percent of back sufferers. But Janet had just spent an hour with a woman who seemed to understand her. Whether the clairvoyant was right about what she told her clients was beside the point. She was right about one thing: Nothing was wrong with Janet's back and there hadn't been for a long time.

Her brain just needed to believe it.

The Illness of Disease

Energy medicine and a rare disease

So we have an explanation. If the disappearance of Adam Engle's back pain was a response to his treatments, there are some sane and science-based—yet still remarkable—reasons why. Adam's disks probably didn't dramatically regenerate or unherniate and his spine didn't go back into optimal alignment, but he did do an impressive number of the things researchers regard as useful for back pain. He made a fundamental shift in how he thought of himself and his pain. Instead of aging, frail, and "degenerating," as his MRIs had suggested, he reimagined his body as strong and resilient. Adam also fully committed himself to his treatment, took responsibility for his healing, and brought to life a belief that things could get better without surgery. He built trusting relationships with attentive healers whose touches sent healing cues to his brain, and he did a lot of exercises without worrying if they were going to destroy his back. He even released a painful memory that had probably dialed up his suffering and perpetuated what might have remained a minor or temporary problem. All of this, if we follow the path science has led us down, is likely to have rewired his brain to conclude that the protective sensation of pain was no longer necessary.

Because, while Adam's pain presumably started in his body— perhaps a knotted-up piriformis muscle and tension in his hips and lower back due to years of sitting in a half lotus position—it was his brain that, unbeknownst to him, reacted in ways that made it chronic. Thus, it was necessary to treat both his mind and relevant areas of his flesh. The work Peter and Norm did with their hands no doubt laid the groundwork for a heavy exhale of tension and irritation from Adam's muscles and connective tissues, which curtailed the danger messages shooting up to his brain. But without altered beliefs, Adam might still be one of the many Americans wrestling with back pain that never seems to go away.

But this is pain, the maddeningly unpredictable sensation that can be swayed in either direction by turns of the mind. Other conditions also cause people to wander, frustrated and desperate, beyond the perimeters of mainstream medicine, and some of them seem at first blush to have little to do with the brain.

Ian was first diagnosed with the rare genetic disorder FOP, which stands for the unwieldy "fibrodysplasia ossificans progressiva," when he was five. A year earlier, his family was living in a Colorado mountain town, and over the span of a few days, young Ian's neck swelled up dramatically and his body burned with a fever of 106. Worried he had gotten a horrible infection, his mom, Amanda, took him to the ER. The attending doctor suspected something far worse and sent Ian to an oncologist. But it wasn't cancer either. On a quest for answers, the family decided to move back to their New Jersey home, where they would have better access to medical care. After much back-and-forth with different specialists in the New York area, the connective-tissue disorder scleredema was identified. Months later, Amanda and her husband, John, connected the dots with the help of another family and the correct diagnosis of FOP was finally given. At the time, only eighty-two cases of it were confirmed in the world. Today, there are some eight hundred, with another thirty-five hundred people estimated to have it.

FOP wreaks havoc on the body in two ways. First, it destroys the muscles, tendons, and ligaments that surround joints. Then it replaces them with bone, gradually imprisoning a person inside a second skeleton. How fast or thoroughly this occurs is unpredictable. Destructive "flare-ups" can be the result of falls or injuries or can just appear out of nowhere. You never know when the next one will occur or which joint it will be. In the seventeenth century, the French physician Guy Patin, who was the first to describe FOP, wrote to a colleague that he "saw a woman today who finally became hard as wood all over." Today, FOP patients don't so much think of their body as hardening as they do of "losing" elbows, knees, jaws, or hips that have become enshrouded in pieces of bone.

When I visit him, Ian is living in a modest redbrick ranch house in a small town in northern New Jersey, not far from the house his family moved back to after the diagnosis. His bedroom is on the first floor, and to get to the basement he uses an elevator that was installed about ten years ago after his hip ossified. He descends into the basement, and I watch him walk in short, robotic bursts across the room. Each step is quickly followed by another, as if any decline in momentum would imperil the whole endeavor. Across his torso, one arm is bent into a fixed pledge of allegiance. The other runs ruler-straight down his side. As he approaches, I realize I am uncertain about whether I am to shake his hand. "Hi, I'm Ian," he says, brushing off my discomfort. He's wearing jeans and a blue T-shirt that drapes loosely over his reed-thin shoulders. His face has a vibrancy his body struggles to match. Without moving his arm, he unfurls his right hand toward me.

I'm here with Ian, now twenty-five, because he's told me that for the past three years he's been getting sessions from one of Donna Eden's trained energy practitioners—sessions, he says, that have helped him more than anything else ever has. "In twenty years, this is the only time I've been able to say that FOP has gotten significantly better," he said the first time we spoke on the phone. "Before this, the best I've had is that it hasn't gotten worse."

Ian has sent me a list of eighteen "body changes" he's noticed since he began working with Gloria McCahill. These include greater muscle definition throughout his body, the ability to lift his arms off his torso by a few more inches than he used to, and the capacity to rotate his left hand so that he can grab the outside of his left thigh. His neck and back, he says, are also "softer and plushier with looser skin"; his legs have less swelling and lumpy growths; his forehead is "smaller and less full"; and he is walking more upright and is able to put more weight on his hamstrings and butt when sitting.

I'm not sure what to make of these claims. They sound impressive, but FOP is hardly the kind of disease you could imagine energy healing or any other alternative treatment being useful for. Not even surgery to remove the unnecessary bone works; a body with FOP greets surgical invasions with the stubborn creation of more bone, making everything worse. Nor are you able to reverse the mutation of the ACVR1 bone-formation gene that causes FOP. The *P* in *FOP* stands for the Latin word for "progressive," which means this disease is supposed to trend in only one direction. It seems foolish to imagine Ian's mind intervening in any of this, tantamount to the brain's trying to regenerate an amputated limb.

Still, I'm intrigued because improvements in FOP are hard to come by and not easily confused with something else. As is often the case with rare diseases, there is a dearth of treatment options. Steroids can help control inflammation, especially if taken during a flare-up, muscle relaxers can ease rigidity, and pain medications will often dull the inevitable discomfort of living within your own cage. But all these solutions only nip around the edges and do so with side effects such as fatigue, lethargy, and bloating. Ian says he hasn't taken pharmaceutical drugs for several years because he doesn't like the way they make him feel.

But energy healing has produced nothing but positive feelings in his body, he says. He realizes that treating FOP with such a thing is deeply unconventional, but he's convinced it's helped, not just because of changes in how his body looks and moves, but because of how he feels daily. "Gloria and I have developed a pretty unique relationship. I'm very open and vocal and let her know what does and doesn't feel good. It's a very fluid, organic interaction."

Gloria is waiting for Ian where she always does, near the massage table. She roots around in her bag, looking for an iPad. Gloria has curly blond hair and pointed features. Wearing a leopard-print scarf, she has an air of warmth and confidence that makes you want to trust her. First trained as a Reiki healer, she has been working with Eden Energy Medicine for ten years. Earlier in the day, before coming to Ian's house, she had given me an "energy session," which Ian is eager to hear about.

"What did you think?" he asks expectantly, while Fitz, his weekday health care aide, helps him up onto the table. "Good" and "relaxing" are the lame words I come up with. In truth, it was much like my session with Peter Churchill in Boulder, though without the intense digging into soft tissues. I felt immensely calm afterward and would do another in a heartbeat, but the effects weren't anything to jump up and down about, although it's true I don't have any current bodily complaints and it was just one session. Ian doesn't seem too disappointed. "It can take time," he says. "I didn't feel that much right away either."

After some small talk, Gloria cues her iPad to soothing flute and piano music and gets down to business. She starts with what Donna Eden calls "clearing the gaits," which means rubbing the spaces between the bones on the tops of Ian's feet and hands and the webbing between his toes and fingers. Then she moves to his leg. She closes her eyes and gently rocks her hands along his thigh for what seems like a long time, as if waiting for it to tell her something. She says she does this to get Ian's leg to relax.

"Neurolymphatic points," Ian explains. "I call them magic points because there are times where my leg gets hard and swollen and Gloria will rub it and right away it starts to soften."

These "magic points," Ian says, are what first convinced him that his weekly or twice weekly, one-to-two-hour sessions with Gloria were doing something. He was in his last semester at the University of Delaware and gripped with the worst flare-up of his life, one that would cost most of the mobility in his second hip joint. His right thigh was swollen to three times its normal size and felt like a slab of concrete. Walking was nearly impossible, and the discomfort was such that Ian couldn't attend classes, so he did his final semester of college from bed.

He went to see two doctors, including Dr. Fred Kaplan, a leading FOP expert, but neither had much to offer outside of pain medications and muscle relaxers, and Ian left despondent. "I thought, 'That can't be it. There has to be something good for me. I can't just be a basket case,'" he recalls, his neck turned sideways on the table as far as it can go, which is just enough for him to look at me.

His aunt, who had been training in Eden Energy Medicine for several years, reached out to offer him a session. Ian was reluctant at first because he doesn't like new people poking around the unfamiliar terrain of his body, even if they're related to him. Often, he says, they're worried about making things worse, and he senses their unease. But he was desperate, so he agreed.

"After a few times, I didn't have any noticeable improvement, nothing amazing, but I did feel something. I remember telling my parents that I didn't know what was going on, but I wanted to continue."

I ask Ian what he means by "something."

"My aunt was rubbing my hands and my feet, and I was feeling a little bit of heat and tingling, which was a new feeling."

Ian's aunt referred him to Gloria, who is more experienced. "In the first few sessions, it was really all about gaining Ian's trust," Gloria says, down at Ian's feet and clasping his big toe between two fingers. "His muscles were in such a habit of fearing anybody coming near him or anybody hurting him. I had to earn his trust not just mentally but in his body. So, I just took my hands and slowly did a kind of gentle rocking on his legs, arms, and back. And I began to notice that when I did that, it started to give and a softening would occur."

Gloria shifts gears. "Give me some trust. Breathe with me," she says, heaving in a huge breath and then pushing it out. Ian shuts his eyes and does the same. For a few minutes, the only sound I hear, apart from the spa music, is their deliberate breathing.

"I know it sounds hard to believe," Gloria says, once they're done, "but I would use a spoon or a fork and just gently stroke certain areas. The first time I did it on Ian's back, he said he got chills and felt a vibration through his body. His back was so hard at that point, I didn't know if he could feel anything. I didn't even know where the spinal column was because of all the hardness. The fact that he could feel through all that gave me a message that this body could be worked on. I just had to listen." She says she would do the stoking of a particular area until the chills and shivers stopped, then move on to another.

"These were places that almost seemed dead to me," Ian adds. "There had been no sensation. You forget what it feels like to feel good in your body when you're constantly in stress or in pain. So, I was restarting a memory of 'Oh, I could have this, this feels good.'"

Gloria is now hovering over Ian's chest and twirling a string with a calcite crystal as a kind of pendulum. "You're out on everything except the third chakra," she informs him. After rebalancing Ian's chakras, she does a "Celtic weave," in which she swoops her hands over Ian's chest in a fluid figure-eight motion. "It really draws all the body's energy systems together in a tight web so that information can go where it needs to," she says, occasionally pausing to shake off her hands.

The session ends with Gloria holding "neurovascular points" on Ian's head in an elaborate series of grasps Donna Eden calls a black pearl sanctuary, a particularly helpful technique, Ian says, that tends to turn him into a puddle. "With FOP, we train our brains to never relax because it's how we survive. We're always on guard and micro-managing and being attentive because you don't know when you're going to fall or somebody's going to bump you or some freaky thing is going to happen, which can result in a flare-up. So, we get used to being tense and don't even know we can relax. Then somebody like

Gloria comes in and shuts my brain off, and I think, 'This is crazy that my body can do that without any drugs or drinks or any outside thing.'"

These neurovascular points, he says, have helped him on several occasions, including once when he was hanging out with a friend a few feet away from where we are now. When he went to pivot on his foot, something didn't go right and he toppled to the ground. "Falling is pretty terrible because I can't do anything to break it. So, I just closed my eyes and took it." He hit his right hip and shoulder and whacked his head on the thankfully carpeted floor.

When Ian's friend got him into a chair, he was ghost white, sweating, and his eyes were rolling back in his head. Both of his parents raced downstairs, and Ian gained enough composure to tell his mom to hold several of the neurovascular black pearl points on his head and do what Donna Eden calls a hookup, one fingertip placed at the middle of the forehead and another at the belly button.

As soon as Amanda did this, Ian felt his body start to relax and come out of shock. He began breathing more normally and the color in his face returned. What could have been a trigger for a cascading inflammatory process, he says, ended up a minor event. "I had a little bit of pain and discomfort in my right hip and in the middle of my back, and I felt pretty fragile for a couple of days. But two days later I went to a wedding in Baltimore with all my friends. I slept in a hotel bed, rode in a car, and did all the stuff that could have aggravated FOP, but didn't."

"That's a big key for healing," Gloria says, pressing her fingers along Ian's cheeks. "You have to feel like there's something you can do."

Just before Ian immerses himself entirely in his relaxed state, he explains that he has begun thinking of FOP in a new way. "Gloria and I think of it more as a superpower than a threat, which makes it less daunting. My body has this crazy ability to hardly ever get sick with colds or infections, to repurpose connective tissue, and to heal injuries really quickly. It just overreacts and does too much of that repurposing sometimes." He says he's also learned to pay far more attention to subtle changes and sensations in his body and to understand what these cues might be trying to tell him. "I'm not so scared of FOP anymore."

Outside, the steady chatter of a cold spring rainstorm lends an air of comfort to the room. Gloria is still holding Ian's resting head, and I use the lull as an opportunity to wander upstairs and talk to Ian's mom, Amanda, who is deeply involved in his care and FOP in general. She attends and organizes conferences and works as a volunteer coordinator for the office of Fred Kaplan, the University of Pennsylvania orthopedist and FOP researcher. I find her behind a pile of papers and books in her home office. "Until recently, I never saw my son's body with no edema, without the swelling and bloating," she says, when I ask her views on the progression of Ian's disease over the last few years. Edema, a common occurrence in FOP patients, is a buildup of the body's lymph fluid. This clear, watery substance helps rid the body of toxins, waste, and other unwanted materials, but unlike blood, it doesn't have its own pump. Its movement relies on the gentle contraction of its channels. FOP's muscle destruction often damages these channels, which can be further impeded by the presence of new bones.

"There were plenty of times when I thought my son didn't have edema," Amanda continues, "but only after working with Gloria for quite a while did I really go, 'Oh, this is what a body looks like without swelling.'"

There are other changes too, she says, beyond the physical. Ian has always tried to maintain a level of independence, staying in college during his leg and hip flare-up to graduate and now working at an app development company he founded, which he does while propped up in bed with his computer suspended above him in an elaborate rig. Recently, he's felt confident enough to want to move out of his parents' house and get his own place, which Amanda says makes her nervous. "It's hard to stop being a mom all the time."

Six months later, when I check in to ask how Ian's relocation went, she tells me she no longer has any trepidation about Ian's not living with her. "It's been a journey for me to realize the psychological empowerment Ian gets from this supportive medicine. He feels very in charge of his body and his care."

Back home, I call Dr. Fred Kaplan. More than a decade ago, he transitioned from a thriving orthopedic surgeon to a researcher of this rare and unforgiving disease FOP. Since then, he and his team have isolated the gene responsible for FOP and have revealed much about the disease's underlying mechanisms. Kaplan has also compiled data on FOP's natural history and met with hundreds of patients, including Ian, whom he knows quite well. Ian calls him "a total gem of a man."

Kaplan tells me he last saw Ian several months ago. "Ian was very pleased and impressed with his progress, as was I. He is certainly a lot better."

Kaplan attributes most of this improvement to two things. First, the inevitable resolution of the intense flare-up that led him to Gloria in the first place. "During a flare-up, the lymphatic channels are blocked. Afterward, they are reconstituted, and this can help patients feel better because the heart isn't working to pump the blood against all this tissue fluid. I've learned that even a small amount of joint movement that's regained when swelling is reduced can have dramatic impact in the quality of someone's life."

Second, there is what Kaplan calls "lymphedema therapy," which, when I look into it, I learn is a special type of gentle massage that coaxes pooled lymph fluid back into its channels, reducing swelling. Four schools in the US provide training in this, often to physical therapists. The most common use of it is to treat edema in cancer patients, as the lymph system is often targeted in an effort to eliminate the cancer.

"I've heard from a lot of my patients in Europe and some in the US who have used lymphedema therapy that it can really help," Kaplan says. "Empirically I can say that the lymphedema therapy in my professional opinion helped make a difference for Ian. Exactly how much of a difference I don't know."

"But," I say, perplexed, "I don't think Gloria has ever been trained as a lymphedema therapist." (I later check, she hasn't.)

"Well, Ian calls it energy therapy, but it incorporates features of lymphedema therapy. He was urinating gallons of fluid from the therapy, so whether Gloria was a trained lymphedema therapist or not, she was clearly doing something beneficial. There was clearly lymph drainage. Both Ian and his mother reported a great deal of fluid excretion. It's not a hocus-pocus thing."

"But how did Gloria know how to do this?" Those who teach lymphedema therapy insist it's not easy to do, you can't just feel your way into it.

"I really don't know."

In my head, I run through Ian's long list of changes—the enhanced muscle tone, the increased mobility at several joints, the softer body parts, the straighter back—and realize that nearly all of them can be explained by the two things Dr. Kaplan mentioned. With less bloat and edema, hidden muscle tissue would be unmasked and joints not fully locked would loosen up, potentially allowing additional increments of flexibility and movement, which could then be followed by strength. This could spur enhanced muscle tone and a greater ease of living in one's body.

Both Gloria and Donna Eden maintain that lymphedema therapy is not the same thing as their "neurolymphatic reflex points," which hail from an early twentieth-century osteopath. But when I look at illustrations of lymphedema massages and Eden Energy Medicine's neurolymphatic points, they clearly overlap. Donna Eden will even tell me that pressing on the neurolymphatic reflex points, though different from lymphedema therapy, can be effective for edema.

Ian insists that there is more to it than this and that recent changes can't be attributed to the easing of a flare-up or the drainage of lymph fluid, both of which are no longer happening. When we talk again months later, he tells me he can make it out onto the deck of his new apartment and use his feet to clear leaves, both actions he couldn't do when he first moved in. I call Dr. Kaplan back to ask whether some of Ian's psychological changes could be responsible for physical ones. Might the greater sense of ease, confidence, and empowerment he feels afford him greater flexibility?

"Who am I to argue with a patient who feels better?" Dr. Kaplan says. "There is more to what a person feels than you can explain by current physiology. There are vast unknowns about the central nervous system and what determines how we 'feel better.' Signals from the brain can alter the resting length of muscles, and you perhaps might be able to get them to reset. Or are these therapies altering the structure and function of the brain to interpret signals coming from the muscles? We don't know."

Dr. Kaplan doesn't say this as an attempt to dismiss the topic. One gets the sense he harbors genuine curiosity for things he doesn't understand. After a long pause, he adds, "One of the things I would like to do is understand what energy therapy really does."

In Boston in late 2000s, researchers at Brigham and Women's Hospital and Harvard Medical School looked at whether a placebo treatment could be of any help for asthma. They enrolled forty-six people with this common disorder of the lungs and asked them to come to the hospital for a series of four different interventions. Two were placebos—a masked inhaler that dispensed a saline solution, and acupuncture done with retractable needles. A third was the actual asthma drug albuterol, which quickly relaxes the muscles surrounding airways and helps these passages open up. The fourth was a visit with doctors where no treatment was given. Patients were blinded as to when their mask had the drug and when it didn't. After everyone had come for twelve visits, researchers discovered that each treatment was equally effective for alleviating sensations of chest tightness, breathing difficulties, coughing, and wheezing. On average, symptoms were 45 to 50 percent better, versus a 21 percent symptom improvement when no treatment was given.

But here was the interesting finding: When pulmonologists had patients blow into a spirometer to see how much air they could quickly exhale from their lungs, there was improvement only when people got albuterol. The acupuncture needles and fake inhalers didn't help anyone's lungs bring in and expel additional amounts of air. Patients felt better—they were less anxious and wheezy and able to breathe more normally—but without any objectively measurable improvement. Earlier such studies had suggested that placebos could help asthma patients take in more air, but none of these had controlled for random fluctuations in lung function or the beneficial effects of being in a clinical trial through the inclusion of a no treatment group.

When the study came out in a 2011 edition of the esteemed *New England Journal of Medicine*, skeptics pounced. Here was definitive proof, they argued, that acupuncture and other placebos did nothing to improve anyone's biology. All they did was trick people into "feeling better," giving them what Dr. David Gorski, a surgical oncologist at the Barbara Ann Karmanos Cancer Institute in Detroit, called medical "beer goggles." Yale's Steven Novella said it was a "game over study" for the placebo effect's reign as a powerful artifact of mind-body healing. Both men seized upon the fact that the study's authors—which included the Harvard placebo researcher and favorite skeptic punching bag Ted Kaptchuk—had appeared to downplay the role of objective tests, such as those conducted with the spirometer. "Even though objective physiological measures are important," Kaptchuk and the other authors had written, "other outcomes such as emergency room visits and quality-of-life metrics may be more clinically relevant to patients and physicians."

More clinically relevant? The skeptics pointed out that, if not treated with medications such as albuterol and inhaled corticosteroids, people can die from asthma. Every year, about a million asthmatic people do just that, mostly in underdeveloped countries where medical care is lacking.

It's a good point. Objective measurements are often a far more relevant measure of the progression of disease than someone's experience of it. Tests can allow doctors to keep people alive. For instance, no oncologist would think of canceling chemotherapy just because a patient arrives to an appointment with a spring in her step. Nor would cardiologists take the disappearance of chest pain or a dramatic shift in mood to be a sign of cleared arteries. But are such subjective accounts of "feeling better" as ephemeral and irrelevant as skeptics make them out to be?

Not if you ask Arthur Kleinman, a noted Harvard psychiatrist and medical anthropologist who, in a series of articles and books starting in the late seventies, drew a distinction between the treatment of *illness* and *disease*. People often use these terms indistinguishably, as I've done to this point, but in Kleinman's view, *disease* is the objective pathology of things you can measure—cancer cells, inflammatory markers in the blood, arterial blocks, blood sugar levels, reduced lung capacity. Such defects, he says, are the primary, if not exclusive, focus of most doctors, who seek to address them with drugs, surgery, the implantation of tiny machines, and the wholesale replacement of organs, sometimes with miraculous results. Disease, Kleinman writes, "deals with the patient as a machine."

Illness, on the other hand, is the lived experience of symptoms, "the monitoring of respiratory wheezes, abdominal cramps, stuffed sinuses or painful joints," and the personal and emotional meaning we attach to such things. It is also the difficulty these symptoms create in our lives and the failure, frustration, anger, demoralization, and depression they breed. "We grieve over lost health, altered body image and dangerously declining self-esteem," Kleinman writes. Indigenous healers and other "folk practitioners," he says, are far more successful at treating these conscious symptoms and their resulting distress than doctors, who "assume that biologic concerns are more basic, 'real,' clinically significant, and interesting than psychologic and sociocultural issues."

The concept of illness helps explain why a person can continue to suffer from a disease but nonetheless declare someone a miracle worker. At the Martinos Center, Vitaly Napadow told me of a woman with fibromyalgia he's been treating for years with acupuncture and who still has fibromyalgia. "To her it's working really well," he said when I asked if his treatments were actually doing anything. "I went away this summer to Korea, and she was ecstatic when I came back." She hasn't been cured, but somehow her time with Napadow makes her feel healed.

Conversely, folk or alternative healers do not often treat the physical distortions of disease, despite what they may profess. According to Kleinman, "only modern health professionals are *potentially* capable of treating both disease and illness." He devised a list of eight questions doctors should ask patients to gain better insight into their illnesses, a proposition many medical schools have eagerly adopted.

Kleinman's thoughts offer, it seems, a useful accounting of Ian's experience with Gloria. In a possibly unusual example of "folk practitioners" treating disease, Gloria's neurolymphatic points/lymphedema therapy alleviated Ian's physical swelling and edema. This didn't cure FOP, but it led to a lot of bodily improvements. Equally successful, if less measurably so, was her treatment of Ian's illness. Despite my initial impressions to the contrary, Ian's mind and brain were critical pieces of his disease. Gloria found ways of talking with him about the fears and frustrations of living with FOP, devised ways of touching that felt enlivening to him, and taught his body how to relax and ease its own tension. He felt supported and understood, and his self-confidence improved and he became more at home in his body.

Such changes in how one experiences day-to-day suffering are hardly unique to rare diseases. Many chronic conditions can inflict psychic as well as physical suffering on their victims, sometimes just by their chronicity. For example, anywhere between 5 to 15 percent of people with multiple sclerosis develop PTSD from the emotional distress of having the disease. "These people can look like they have a huge disability due to MS, but much of the disability may be due to the PTSD," a neurologist told me. "This clinical situation is often underrecognized and underdiagnosed." The incidence of major depression in cancer patients can be as high as 38 percent.

When I reach Kleinman, now seventy-seven and still on the faculty at Harvard Medical School, he tells me that he regards the concepts of illness and disease as less useful than he once did because of how his ideas have been adopted by the medical community. Instead of using his eight questions to build caring relationships and to practice being present with patients, many doctors, he says, have employed the queries robotically, turning complex, vivid lives into "limiting and biased cognitive stereotypes."

"For the most part, doctors haven't been given the time or training to really know how to understand patients, and even if there is training, it gets lost and forgotten amid everything else doctors have to do," he says. The distinction between illness and disease is also, he now thinks, blurrier than he once considered. Improvements in a disease will, after all, improve one's experience of illness, and conversely he thinks alleviations in one's illness can reverberate into bodily shifts of disease.

"Like how calming the mind can relax and maybe reset the resting length of the muscles?" I say, referencing Dr. Kaplan's comment.

"There are all kinds of possibilities. We know that shifts in the way someone thinks and feels can slow heart rate and lower blood pressure, and there are probably lots more. I don't think we know the extent of it."

Back when I was in San Diego, Fabrizio Benedetti had said there wasn't any evidence a placebo effect could reverse cancer, heart disease, diabetes, osteoporosis, or Alzheimer's. In the early 2000s, he had tried and failed to get a person's expectations to do for hormone secretion what they've been shown to do for pain. He had injected people with saline and told them it was a powerful drug that would stimulate their growth hormone, but he got no response. He told another group it would decrease their cortisol, the so-called stress hormone. But again, nobody's pituitary or adrenal gland increased or decreased anything. Manfred Schedlowski, who runs the Institute of Medical Psychology and Behavioral Immunobiology at Germany's University of Duisburg-Essen, tried something similar with the immune system, giving people "a powerful immunosuppressant drug" that promised to decrease immune signaling proteins such as interleukin-2. Once more, nothing.

Yet many conditions *are* known to be responsive to our positive expectations about a treatment and meaningful interactions with the person providing it: anxiety, depression, nausea and other gastrointestinal distress, sleeplessness, the shortness-of-breath sensation that comes with asthma, fatigue, itch, and, of course, pain. But why is a placebo good for some health concerns and not others? Surely its list of achievements isn't happenstance.

The answer seems to lie in the nature of things we know placebos can be effective for. Like pain, each of these conditions is created largely within the flurry of chemical and electrical activity between our ears. Pain, as we've seen, is the brain's evaluation of danger. Nausea feels as if it's happening in our gut, but the sensation arises in a nausea center in the brain stem, often in response to messages from the stomach and intestines. Similarly, feelings of fatigue are guided by a variety of bodily reports—heat, hydration, blood acidity levels, and the presence of immune proteins called cytokines, which are responsible for that lethargic blah sensation of being sick. But ultimately, as scientists are learning, fatigue is another of the brain's protective perceptions. Instead of changing some area of the body, placebos and other interventions that target the mind are altering our brain's governance of such felt experiences, shifting its evaluation of whether to evoke a particular symptom. Benedetti had said "care, not cure," but he may as well have said illness, not disease. This is why a placebo can take away pain but not heal a torn muscle or repair damage to nerves; change how your lungs feel but not actually widen their passages; temporarily increase dopamine in the brain but not prevent the death of further dopamine-producing cells; quell nausea but not get rid of a stomach virus.

During the incipient stages of this book, I thought about my husband's Stargardt disease, which is an inherited form of macular degeneration that can't be corrected with glasses, laser surgery, or anything else. He was born with this genetic disorder, though it didn't show up until he was in college. I figured that the possibility that a mind-body treatment could somehow help him regain some of his lost central vision was a shot so long even a perfectly sighted person couldn't see it, but I was doing an exercise in what-if, so my imagination wandered. I now realize that no pathways are known by which my husband's mental activity, no matter how trippy or transformative, could clear the fatty deposits in his retina or regenerate his dead photoreceptor cells. It's just not something you can get to that way.

Nor can changed brain activity resolve the silent acid reflux I now know to be the cause of my throat problems. Several months after Peter Churchill surmised I would be much better, I had one of those horrible episodes in which I felt as if my throat were clamped shut, followed by another a few months later. This prompted me to reconsider what that gastroenterologist had suggested years ago. I started to reduce or eliminate many of the foods and beverages implicated in acid reflux—sadly, coffee, wine, chocolate, and pizza sauce. After abstaining for two months (on everything except coffee; I could only bring myself to cut back), I noticed a significant improvement in the number of times I felt that tense, tugging sensation while eating. I have since reintroduced these items at lower volumes and find that if I stick to less than desired quantities, I hardly ever feel the constriction.

More significant, I haven't had a severe episode since I accepted the reflux theory, which was two years ago now. A visit to a new gastroenterologist confirmed that I have a small, herniated opening in the valve that keeps stomach acid from heading north. This, combined with acidic foods going down, causes irritation of the esophagus, sometimes leading to muscle contractions and the closed-up feeling. I do think Peter was right that there's fear and hypervigilance at play, and that my mental state and stress levels can affect how reactive my esophagus will be to these acidic assaults. Yet the primary driver for the whole thing is clearly biochemical, not neck tightness or my kids fighting or my remembering how bad it was last time. I could spend months Zen-ing myself out and it wouldn't stop the acid from coming up and triggering those muscle contractions.

As the body's command center, our brain is linked to every major system in the body. There are pathways to the cardiovascular system, the immune system, the hormones of the endocrine system, the muscular system, the sexual system, and so on. The million-dollar question is how much we can use our minds to gain access to all this for healing. In fascinating results that suggest both the limits and possibilities of mind-body interactions, placebo researchers have shown that we can stimulate our hormones and alter aspects of our immune system. The catch is that we can't do it on our own; we need a drug to help us hack into the unconscious parts of our brain. In part two of those placebo experiments that Benedetti and Schedlowski did on hormone secretion and immune responses, people were first given the actual hormone-altering or immune-suppressing drug for several days along with a "memorable stimulus," such as a sniff of peppermint oil or spoonful of anise-flavored syrup. (Benedetti used the drug sumatriptan and Schedlowski the immunosuppressant cyclosporin.) Then after four or five days, the drug was surreptitiously swapped out for a sugar pill or saline solution, still paired with the stimulating syrup. And voilà, a placebo response. Measurements of growth hormone went up or cortisol went down, and interleukin-2 decreased. In other studies, researchers showed that you could do similar things with the secretion of insulin, and more recently Benedetti found that fake oxygen could tamp down the body's circulatory and breathing responses to the absence of oxygen at high altitude. Yet to do this, you needed to give people real oxygen first. A bioactive molecule, whether oxygen or a drug, was necessary to train the brain to activate the unconscious signaling systems that exist for hormone secretion, immune responses, and circulatory changes—pathways that conscious thoughts and beliefs were useless for.*

---

*This represents another possible honest use of placebos in medicine. Drugs could be given for a few days, then swapped for sugar pills, and then back to the drugs and so on. But since this placebo response has "extinctions," you would always need to keep going back to the drugs.

This wasn't the placebo response as everybody had come to think of it. In fact, when Benedetti told patients who were taking a placebo growth hormone that their levels would go down, they went defiantly up, mirroring the effect of the actual drug they'd taken a few days earlier. "The working hypothesis," Manfred Schedlowski says, "is that these autonomic systems like the immune system and the neuroendocrine system can be modified mainly by these associated learning or conditioning procedures. Whereas pain perception and other symptoms are conscious events that can be modulated by expectations and patient-doctor communication."

This doesn't mean, however, that any drug can condition a Pavlovian placebo response. For instance, no brain-body pathways are known for chemotherapy drugs, antibiotics, or statin medications. On some level, the body *is* a machine, as horrible and Cartesian as that sounds. There is both profound holistic symmetry and an inherent organizational structure with rules and boundaries. As much as we might like to believe otherwise, we don't have control over many of the processes taking place within our bodies or inside our brains. At least not using beliefs, expectations, or intentions. My next question was whether other mental activities such as relaxation and meditation could take us to places these other things couldn't. Could they engineer effects beyond the brain, in such realms as the immune and endocrine systems, which we now know to be deeply intertwined with our nervous system?

10
. . . .

The Zen Response

Stress reduction and the immune system

The oversize grape rolling around in my hand suddenly feels like a rounded piece of lead. During a bathroom break, our instructor had surreptitiously stashed them under our chairs, just a single grape in a small paper cup. "Take your grape in your hand and really focus your attention on it," she had told us. "Get curious and realize that everything is new: What does the grape look like? How does it feel between your fingers? Is there a smell?" I've been doing my best to zoom in on the grape-centric moment and am now surprised to find that, amid the quiet stillness of the room, this garden-variety fruit holds a few surprises. Along with the unexpected heaviness is a perfect and pearly smoothness and, as the grape is rolling around unchewed in my mouth, an explosive sweet-tart taste from the tiny stream of juice that has oozed out.

"Some of you might be loving this exercise or maybe you're thinking it's ridiculous," says the instructor, Gloria Kamler. "The judgey part of the mind always comes up. We just quietly notice—whichever it is, it's fine." A jocular New Jersey native, Gloria has been taking and teaching classes like this one for the past twenty-five years. Their subject is mindfulness, a state of focused attention to the present moment that often uses meditation. By her own estimation, this six-week-long class held at UCLA's Mindful Awareness Research Center

(MARC) is at least the one hundredth she's taught, here at UCLA and elsewhere.

When we finally chew and swallow our grapes, she asks for words to describe our experience. Lots of people say "sweet," "round," and "smooth." Someone says "skeptical," which Gloria likes. "I really want you to notice what it's like for you to focus on your experience of the present moment," she says. "The habit is always to go somewhere else, but mindfulness is about how we keep showing up and letting ourselves be surprised." I settle on "magnified" because the exercise seemed to have tuned up my senses.

When the molecular biologist and yoga teacher Jon Kabat-Zinn first devised this activity in the 1980s for his "mindfulness-based stress reduction" classes in Boston, he used raisins. His logic was that getting people to focus on ordinary, everyday sensations such as eating would give them insights into what could happen when their mind divorced itself from the worries of the future and the dramas of the past. This and other Kabat-Zinn techniques have been adapted for classes taught here at MARC. They take place pretty much year-round and are open to anyone who wants to participate, and if tonight's class is any guide, that would seem to be everybody. The twenty people assembled here on a Wednesday night in a room inside UCLA's neuroscience building are a demographic smorgasbord: a number of grad students in their early twenties, a few people who look to be in their sixties, and many in between. The number of men almost matches the number of women, which is unusual for anything to do with self-help or that makes use of the word *emotion*. Unlike the Cleveland Clinic pain class, which was all-white and mostly women, our group is white, black, Indian, Latino, Chinese, and Middle Eastern. Two women explain that they have come because they have chronic pain that doctors can't identify a source for. Others say they need help managing stress and anxiety in their lives. Two men say they already meditate but aren't sure they're "doing it right."

I can relate. The first time I did meditation, in the early 2000s, was during a period of my fascination with Buddhism, the religious tradition from which this practice hails. I was living in New York City and frequenting a Buddhist center in midtown. I'm sure I attended some interesting talks and discussions, but the experiences I remember most vividly are the many lengthy stretches of meditation we did. Sometimes these were achingly long, an hour or two, and offered a revealing introduction to the hyperactive five-year-old that lives inside the human mind (or at least does in mine) and the difficulty one faces in getting her to shut up and pay attention. I had many moments of deep equanimity when she did finally stop babbling and worrying, when it felt good to just be quiet and still, but the lessons never sank in. I understood that the idea was to develop a more peaceful, accepting approach to life, to be the sort of human who would notice the people yapping away on their cell phones in the train's quiet car but not give them the stink eye or privately stew about it. But I suppose I never cared enough about becoming this person, and meditation always felt like something I should do instead of something I actually wanted to do. I also, much like Vitaly Napadow, the Harvard researcher and acupuncturist, found it hard to stay physically still for so long. All I wanted to do was get up and wring out my creaky joints. It's true, though, that I never had a teacher like Gloria, who laughs a lot and seems to remember how strange and impossible learning how to meditate is. "In my early years, I spent a lot of time sleeping," she tells us, illustrating the head nod she calls the "Zen jerk."

Gloria explains that this class is all about developing life skills, not learning a religious tradition. She says that by getting to know one's mind and training it to be less reactive to life's constant barrages of stress and negative emotions, people can dramatically change their lives and improve their health. "Studies have shown that when we deal with stress in unhelpful ways, like burying it or fighting against it or letting it control us, this can lead to all kinds of mental, emotional, and physical problems," she says, articulating a view that used to prompt controversy but is now more or less accepted scientific truth.

Research done over the past three decades has repeatedly shown that in addition to making us crabby and irritable, stress can cause depression, create anxiety and sleeping difficulties, increase pain, sap energy, and promote muscle tension. Even more troubling, it can encourage inflammation and throw our immune systems off-kilter, which may contribute to the development or exacerbation of such problems as heart disease, type 2 diabetes, and autoimmune disorders such as rheumatoid arthritis, multiple sclerosis, and psoriasis.

That this can happen—that stress stewing in our heads can burrow its way deep into the organs and tissues down below—has everything to do with the evolutionary fight-or-flight response that's part of our sympathetic nervous system. When we feel intense fear or panic, our brain's amygdala sets off a rapid-fire chain of nerve signals to the hormone-secreting glands of the body's endocrine system. First, the messages travel in the brain to the nearby hypothalamus, then down along nerves to the adrenal glands, which are perched atop the kidneys like hats. There, the hormones (and neurotransmitters) adrenaline and noradrenaline are released, which sets into motion a series of events that get our body ready to fight or to flee. Our hearts beat faster and our blood pressure escalates to send more blood to the arms and legs. Breathing becomes more rapid so we can take in more oxygen, some of which gets sent to the brain to increase alertness. Extra glucose and fat are released for energy. It's a handy mechanism for short-term situations in which we need to act quickly to avoid getting mauled or run over. The trouble occurs when stress is the perpetual, non-life-threatening variety. Low-level fear, anxiety, frustration, worry, anger, and other adverse emotions also activate this sympathetic nervous system, albeit to a lesser extent, resulting in continuously elevated levels of hormones like adrenaline, noradrenaline, and cortisol.

This all sounds worrisome, and it is, but it wouldn't be nearly so troubling if not for what was discovered in the early eighties. Immunologists used to think that the body's immune system for repairing itself and fighting off infection—its many white blood cells such as T-cells, B-cells, natural killer cells, neutrophils, and macrophages—

was a self-contained unit. Then in 1980, scientists working with mice at the Indiana University School of Medicine found a network of nerves inside glands that led to cells of the immune system. Not long after, University of Texas researchers revealed that human immune cells were capable of producing those same stress hormones that are generated by the pituitary and adrenal glands. At first, some scientists insisted these results must have been a function of dirty test tubes—white blood cells of the immune system were supposed to be distinct from the hormones of the endocrine system. But a few years later, others replicated the findings and discovered that not only did immune cells produce adrenaline and noradrenaline, they also had receptors for receiving them. This meant that the nervous, immune, and endocrine systems were intimately and dynamically linked, which was big news and spawned the new field of psychoneuroimmunology. It also meant that stress had a pathway to meddle in immune responses, the reason being, some have theorized, that in prehistoric times acute bursts of stress often meant the likelihood of an injury and thus our immune systems needed to be ready.

It's now thought that stress, whether temporary or ongoing, affects immune responses in two opposing ways. One is by suppressing the type of response that allows us to fight off foreign invaders such as viruses and bacteria. The other is by activating immune responses involved in the repair of damaged or infected tissues. When some part of the body is injured, white blood cells flood the area and release molecules such as cytokines, which increase blood flow and help clear out invaders and dead cells, in a process otherwise known as inflammation. We think of inflammation as troublesome—there's redness, swelling, and pain—but it's a price to pay for healing and recovery. This restorative process does turn destructive, however, when stuck in the "on" position because of a continuous fight-or-flight stress response. Chronic inflammation is believed to be a contributing factor in all those health problems mentioned above such as heart disease, type 2 diabetes, and autoimmune diseases. (Inflammation does, it should be noted, have other causes besides stress.)

This is all the bad news. The question scientists who study psychoneuroimmunology at UCLA have been grappling with is whether
techniques that reduce stress, such as mindfulness meditation, can
move things in the opposite direction. If stress creates this dysregu-
lated, inflammatory turmoil, can relaxation lead us out of it?

The first time Michael Irwin saw a tai chi class, he was not impressed.
It was 1993 and he was at a Kaiser Permanente facility in San Diego
where classes were being offered to members. A bunch of older people were trying to do these very, very slow movements. They'd languidly raise their hands over their head before bringing them back
down, then very deliberately push their hands out in front of them
as though they were kneading imaginary large balls of dough. Most
had significant disabilities and were overweight. Many were doing
the exercises in a chair because they couldn't stand for any length of
time. *This isn't going to work*, Irwin remembers thinking. *These people
have serious problems and this kind of stuff isn't going to change that.* At
the time, Irwin was a psychiatry professor at UC San Diego's medical school and had gone to observe the class with one of his students
who was doing a research project on exercise.

When Irwin went back to the class at the end of six weeks, he was
surprised to find the chairs gone. Everyone was now able to stand
while doing the forty-five-minute class. "I thought, 'Wow, there's
something happening here,'" he recalls. He decided to try tai chi, the
practice of slow, focused movements that some call "moving meditation," for himself and ended up practicing it daily for many years. "I
found that it really gave me more energy and greater sense of balance
and strength."

Irwin is telling me this while sitting legs crossed in his ample
office overlooking the Ronald Reagan UCLA Medical Center. He has
white hair and a trim white beard and speaks in the kind of calm,
measured voice that suggests he's learned a thing or two about relaxation. In addition to practicing tai chi, which he now does sporadi-

cally, he does yoga and meditation and attempts to bring mindfulness into his daily activities. "On my way over here this morning, I went to the gym first, and Google told me to go a different way than I normally do, so I had to be mindful not to start screaming at the map lady," he says, revealing himself to be much like every other person in LA.

Irwin grew up in Wyoming, but is a longtime LA resident. He was here at UCLA for medical school and his psychiatry residency before moving down to San Diego and living there for a decade. He returned in 2001 to become the director of the Mindful Awareness Research Center and UCLA's Cousins Center for Psychoneuroimmunology. While many of his colleagues in the psychoneuroimmunology field have strived to uncover negative associations between stress, the brain, and the immune system and to find pharmaceutical solutions to interrupt them, this has never interested Irwin. For the last twenty-five years, he's been asking a different question: Does reducing stress through practices such as mindfulness meditation, tai chi, and yoga improve one's health? Does it regulate the immune system, reduce inflammation, and relieve symptoms?

One of the first studies Irwin performed when he got to the Cousins Center was on tai chi and immune suppression in older adults. He randomized thirty-six people to get either fifteen weeks of three weekly tai chi classes or be on a wait list, and found that doing tai chi improved physical functioning and boosted biochemical measures of immunity for the varicella zoster virus, which causes shingles. He replicated the findings a few years later with a larger study that showed doing tai chi and getting the vaccine for the varicella zoster virus gave a bigger boost to immunity than just the vaccine by itself.

Many of his colleagues in the field were skeptical. Some had no idea what tai chi was, and the number of people who had at the time done randomized clinical trials on the practice could be counted on two hands. And those studies had looked at such things as physical balance and muscle function, which you could logically see such exercises being potentially helpful for. Irwin was investigating something else entirely.

"In the early 2000s, I organized a symposium on mind-body interventions and health and presented our data on tai chi and the shingles immunity. But even before the presentation began in earnest, many people just walked out of the room, complaining that such research was mumbo jumbo. It was kind of traumatic to me as a scientist. Some of my colleagues told me I had lost my grounding as a scientist to study tai chi."

Rattled but still confident in the rigor of his research, Irwin forged ahead with more tai chi studies. He obtained funding from the National Institutes of Health to look at whether the practice could help improve symptoms such as insomnia and depression, which a growing number of studies link with inflammation. For insomnia, inflammation is a downstream effect of not sleeping well, and for some instances of depression, it could be a contributing cause. In a 2008 study, Irwin showed that older adults with moderate sleep problems who did sixteen weeks of tai chi saw significant improvements in their quality of sleep as compared to those attending sixteen educational presentations about good sleep habits, healthy nutrition, and the health benefits of exercise and relaxation. A more recent study, published in 2017, showed that tai chi for sleep problems in breast cancer survivors was almost as effective as insomnia-specific cognitive behavioral therapy, or CBT-I, which is considered the best nonpharmacological treatment for sleep problems. Thirty-eight percent of people getting tai chi and 46 percent getting CBT-I were completely freed of their sleep issues, though the effects from tai chi instruction, which wasn't specifically tailored for insomnia, lasted only while people were taking classes in it.

"It's pretty amazing that tai chi could be as good as the gold standard, especially since this wasn't some specific intervention about sleep the way CBT-I is," Irwin notes.

In these and other studies on people with insomnia, Irwin showed that practicing tai chi made people calmer and reduced their stress response, as measured by heart rate variability. It also significantly reduced the ability of immune cells to produce those inflammatory cytokines in response to stress. Finally, it decreased the expression of genes that signal inflammation. All these effects were seen both in patients who reported improvement in their insomnia and in those who didn't, whereas CBT-I appeared to reduce inflammation only if insomnia improved. What this suggested was that CBT-I had an impact on inflammation only as a downstream effect of better sleep, but that tai chi could reduce inflammation by alleviating someone's insomnia symptoms, by lowering their symptoms of stress, or both. (Studies done elsewhere have shown that tai chi can have positive effects for pain, notably back pain, knee osteoarthritis, and fibromyalgia, and that it can boost brain function and reasoning ability in older people.)

Irwin and his team of fifteen UCLA faculty members have also done studies showing that mindfulness meditation classes like the one I attended can lead to similar reductions in the stress response, insomnia, and depression, as well as a down-regulation of those biomarkers for inflammation and immune activation. "Now fellow scientists go, 'That's great stuff, how interesting,'" Irwin says. "It's striking to me that scientific meetings that once scorned research in mind-body medicine not only routinely highlight such research but even offer meditation, yoga, and tai chi classes for attendees. I feel like people in the biomedical community are beginning to see the value of these approaches, both from a public and personal health perspective."

But what is that value exactly? It's impressive that Cousins Center studies have shown benefits for depression, anxiety, sleeplessness, and fatigue, given that nearly all of us will succumb to one or more of these things in our lifetimes. And the reductions in inflammatory markers and gene expression would seem to be hard evidence that a mind-body intervention can positively change a meaningful aspect of the body's biochemistry. But Irwin's team hasn't tried to see if these changes can do something clinically remarkable such as reverse major diseases where inflammation plays a leading role, such as heart disease or type 2 diabetes. Nor have they shown that meditation, yoga, or tai chi can turn off immune responses that have gotten inexplicably stuck in the "on" position, such as in autoimmune diseases. In these disorders, the body is essentially attacking itself in some particular area: the joints of the arms and legs (rheumatoid arthritis), the lining of the intestines (inflammatory bowel disease and Crohn's disease), the connections between nerve cells in the brain (multiple sclerosis), the insulin-producing cells in the pancreas (type 1 diabetes), the nerves controlling muscles in the legs and sometimes the arms and upper body (Guillain-Barré syndrome), the skin (psoriasis), the thyroid gland (Graves' or Hashimoto's disease), and the joints and connective tissue of the spine (ankylosing spondylitis).

"That's the million-dollar question," Irwin says. "If you have a disease population, can you put into place these measures that will change the disease?"

His answer is, not necessarily. "I think that once you've got a disease process in play with these underlying pathophysiological mechanisms, it's very hard to reverse it. The ability for an intervention like tai chi or mindfulness may not be strong enough or robust enough to overcome that."

A decade ago, the Cousins Center did a study with 130 people who were suffering from rheumatoid arthritis. Half of the participants were given tai chi and the other half cognitive behavioral therapy. The results dumbfounded Irwin. "I would have expected that tai chi would have reduced the cellular production of pro-inflammatory cytokines, like we saw in breast cancer survivors with fatigue and older adults with insomnia. But it just didn't."

In the CBT group, some modest effects on pain, inflammation, and immune markers were seen. In the tai chi group, there was nothing, no down-regulation of inflammation or immune response whatsoever. It didn't even help for pain. These patients were still very much living with rheumatoid arthritis. The study was never published because the lead author didn't know how to explain the results, but lately Irwin's been itching to get it out. "When people don't get the data they like, they say, 'Well, I'm not going to write this up.' But I think we should report the data and say these stress-reducing treatments like tai chi could be effective for different symptoms if you don't have an active inflammatory disorder."

Such as ankylosing spondylitis. The Cousins Center is named after Norman Cousins, who was an author and peace advocate. He became interested in mind-body healing after being diagnosed with ankylosing spondylitis, the autoimmune disorder that attacks the joints and connective tissues of the spine. In his book *Anatomy of an Illness as Perceived by the Patient*, Cousins describes returning from a trip to Russia in 1964 with extreme malaise and achiness that intensified into an inability to move his neck, arms, fingers, and legs. After receiving his ankylosing spondylitis diagnosis from a doctor, Cousins decided to try to heal himself through laughter. He watched Marx Brothers films and episodes of *Candid Camera*, first noticing that deep belly laughs had a temporary pain-relieving effect that afforded him restful nights of sleep. Then he started to feel more of his symptoms subsiding. Several months later, he says, all of his sickness was gone, fueling a deep curiosity about how such a thing was possible. The book went on to become a bestseller and was made into a 1984 movie starring Ed Asner. Cousins forged a relationship with UCLA and helped put together a psychoneuroimmunology task force that was later expanded into the Cousins Center.

In the years since, many people have expressed skepticism about whether Cousins had ankylosing spondylitis. The disease has no cure and doesn't typically go away on its own, though it does have temporary periods of remission. Some have suggested that he had reactive arthritis or polymyalgia rheumatica, both of which can produce similar symptoms but often clear up on their own.

Irwin says he doesn't know what exactly Cousins had for those few months in 1964, but says it probably wasn't ankylosing spondylitis. Nor does Irwin think it's possible to laugh one's way out of such a significant physical malady. "Oh, no, no," he says, nearly laughing himself, "I have never promulgated that notion, that you laugh and recover from diseases. What Norman had is just not consistent with ankylosing spondylitis. Although we and others have repeatedly found that an optimistic attitude may have a protective role in preventing disease in the first place."

I ask Irwin if it makes him uncomfortable that his center is named after someone whose most famous claim about healing is probably incorrect. "No," he says without hesitation. "I see his message as so much broader than that. Norman challenged the biomedical community to consider the whole person in the healing process, and to not limit our understanding of disease and its treatment to simply giving a pill or some other unitary treatment. He emphasized the humanism of medicine and the importance of forming a partnership between the patient and the provider. Additionally, he recognized the role that each of us can play in our health and healing, and he understood that finding meaning in an illness, even a cancer diagnosis, predicts a more favorable outcome."

Instead of reversing autoimmune or other inflammatory diseases, Irwin thinks it's possible that stress-reducing therapies could help pump the brakes on the advance of some of these diseases, especially if you do them early and stick with them. A few years ago, a study done at Northwestern University's Feinberg School of Medicine, for example, showed that after sixteen one-on-one sessions of cognitive-behavioral stress-management therapy, people with multiple sclerosis had fewer new brain lesions on an MRI than those participants on a wait list, but these effects lasted only as long as the patients were getting the therapy.

Some types of dysfunctional immune reactions may also be easier to turn off than others. Immunologists have different classifications for immune hypersensitivity, with rheumatoid arthritis and multiple sclerosis being among the more complex. Allergies, on the other hand, are relatively simple responses to environmental irritants, which is perhaps why they can respond to placebos and acupuncture. In addition to the tactile cues and attentive rituals of acupuncture, many patients feel their thirty- to sixty-minute sessions have a stress-relieving component. Yet as the Maryland placebo researcher Luana Colloca pointed out after her allergy symptoms disappeared, the effects of expectation, relaxation, and other mental characteristics on allergies aren't well studied.

Also riddled with uncertainty are questions of stress and cancer. After leaving Irwin's office, I make my way to the other side of the UCLA campus where Julie Bower, who has been leading the Cousins Center's trials of cancer survivors, inhabits a fifth-floor perch in a rectangular block of cement known as Psychology Tower. When I knock on her door, she's in the middle of assembling a syllabus for an ambitious undergraduate class she's teaching called Health and Mind-Body Relationships. Bower has long blond hair, a chirpy manner, and, like Irwin, a background in psychology.

Much of her work has been with women who've had their breast cancer successfully treated but then experienced lingering problems such as insomnia, depression, and fatigue. She's shown that yoga and mindfulness meditation can help with these, perhaps because they are toning down inflammatory activity. "Some people just have this horrible nagging fatigue that can last for years," she says. "It's about twenty to thirty percent of breast cancer survivors, and we think that what's going on is that there's this inflammatory component of the immune system that's getting turned on in response to the cancer treatment or the stress of a cancer diagnosis and isn't turning off." Bower thinks that either someone's body is predisposed to mount a stronger-than-normal immune response, or her brain is more sensitive to those pro-inflammatory cytokines released by immune cells. In addition to creating inflammation in the body, these molecules also produce that feeling of malaise and depletion you get when you're sick with the flu or an infection. They work by signaling the brain to produce similar molecules, generating exhaustion so you'll stay in bed and rest. If the brain is producing such chemicals when you're not actually sick, a feeling of inexplicable fatigue sets in.

But as for helping people shrink their tumors, that's not exactly on the agenda here at the Cousins Center. "Meditation gets a lot of attention these days, but it's hardly a cure-all," she says. A big study UCLA is doing with the Dana-Farber Cancer Institute and Johns Hopkins Medicine, for example, looks at how mindfulness classes can help young breast cancer survivors with stress, depression, and fatigue—not tumors. Bower does wonder, though, if stress plays a role in the ability of cancerous tumors to return or to spread to other areas of the body, since certain types of stress-responsive immune cells are known to be active in helping tumor cells escape into the blood and lymphatic system. "We have a whole program of basic research here looking at links between stress and tumor biology."

She mentions a study done in Israel that gave thirty-eight patients with active breast cancer multiple doses of beta-blockers, which thwart adrenaline's effects, and COX-2 inhibitors, which reduce inflammation. When these patients had their tumors removed and analyzed, their cancer cells looked less metastatic than the cells of those who were given a placebo. These researchers also got similar results in patients with colon cancer, and trials are under way with other types of cancer. Such research, though preliminary, suggests that toning down activity in the sympathetic nervous system may be a promising approach to reducing inflammation and perhaps making cancer less likely to spread. But we shouldn't leap to any soaring conclusions. "It's a proof of concept, but we still don't really have enough data yet to know what it means clinically," Bower says.

I leave UCLA with the conviction that were I to get significantly depressed, anxious, sleepless, or tired for no apparent reason, I should find myself a mindfulness class taught by someone like Gloria Kamler. Maybe I'd also consider doing tai chi or a mindful type of yoga. But if I got an autoimmune disorder or a cancer diagnosis, I shouldn't harbor the expectation that learning how to calm my mind or my body, or some synergistic combination thereof, could cure me. I would remind myself, though, that learning to better handle mental and emotional strains, now presumably arriving with greater frequency and intensity, could help control my symptoms and deliver the sort of peace and acceptance that can be its own kind of profound healing. Maybe I'd also cautiously hope that whatever practice I was doing could slow down or stabilize my disease. Because Irwin's and Bower's research does suggest that improving the chaos in one's mind can reach down into the body to improve inflammation. But there's not much evidence for anyone's being able to om his or her way out of a disease actively ravaging the body. As far as science is concerned, there are still no mind-over-matter miracles. Though sometimes it sure can look that way.

The Emotional Rescue

A miracle at Lourdes

It all started in the middle of a large and craggy rock. On a cold February morning in 1858, fourteen-year-old Bernadette Soubirous was sent out by her mother to collect firewood for cooking. She headed out of the village with a friend and one of her four younger sisters to an area where pigs and cattle grazed and where, in warmer months, children came to bathe or fish for trout. This wild and sometimes rancid place, ringed by a river, was covered in what one account describes as "blood and pig hair." Bernadette's sister and friend waded firmly through the frigid river, but Bernadette, who suffered from asthma and had never fully recovered from a bout of cholera, sat down in an attempt to preserve her dry socks.

As she was removing her stockings, she noticed it. On the other side of the river, where a large dark outcropping of rock folded inward to form a shallow cave, was a child about Bernadette's age dressed in a flowing white robe and blue scarf. She was standing in one of the rock's indents just above the cave's opening, next to a wild rosebush. Barefoot and bathed in soft white light, she looked calm and peaceful. As she stared at the girl, Bernadette heard a stiff wind suddenly blow, although none of the trees or bushes seemed to move. Startled, she padded her hand around on the ground to find her rosary, which she often had with her. But Bernadette was too frightened to pick it up. She watched as the figure produced its own rosary and made the Catholic sign of the cross. Then, as mysteriously as she had appeared, the girl disappeared. Bernadette quickly scrambled across the river to her companions. "You saw it, didn't you?" she asked. Her sister and friend stared at her blankly. They hadn't seen anything. Nor had they heard any wind. Whatever it had been was only apparent to Bernadette.

At home, Bernadette's sister explained the incident to their mother, who grew angry and forbade her daughters from ever going back. But three days later the girls went anyway, this time with another sister, more friends, and holy water from a local church. Again, the image appeared only to Bernadette. To test whether it was an evil spirit, she threw the holy water at it, to which the glowing girl gently smiled and bowed her head until the bottle was empty. One of Bernadette's friends had another test in mind. She threw a large rock at the area where Bernadette said the invisible girl was, to see if that would knock her off her perch. Everyone except Bernadette jumped as the rock crashed against the much bigger one. Deeply pale and immobile, she looked as if she were paralyzed, in a trance.

On the third visit, now with dozens of curious onlookers in tow, Bernadette tried to speak to the ghost, which was still standing in the same spot. To her surprise, it spoke back: "Would you have the goodness to come here for fifteen days?" So for fifteen more days, Bernadette visited the grotto, as it's now called, with an ever-increasing crowd of villagers from the small town of Lourdes. It was quite a show. Newspaper reports noted how Bernadette's "hands begin to tremble and the nervous twitching sets in." Her lips, they said, "shook convulsively." One person went so far as to poke Bernadette's shoulder with a sharp pin while she was in one of her states, eliciting no reaction at all. On the seventh visit, the apparition revealed her identity to Bernadette: She was the Immaculate Conception, or Mary, the mother of Jesus, a proclamation that created quite a stir in the heavily Catholic town. Many villagers believed Bernadette and were awed that such a divine figure would appear in their town.

A few, though, remained skeptical, notably the chief of police and a prominent local priest, both of whom suspected Bernadette was orchestrating an elaborate teenage hoax. Why would Mary the Virgin Mother appear to a poor, sickly child in the middle of nowhere? Both men would later change their minds about Bernadette after she appeared to have no interest in the fame or attention her experiences afforded her. The famous teenager left Lourdes not long after the ordeal to join a convent, where she quietly lived out the remaining twenty years of her life.

On the ninth visitation (there would be eighteen in all), the Virgin Mary told Bernadette to dig for an underground spring inside the cave. Anyone who wanted forgiveness could drink from this spring or wash in it, Mary said. When Bernadette did this, a muddy pool of water emerged that, according to the Church's official telling, eventually became a clear spring. A few days later, a pregnant woman named Catherine Latapie arrived at the spring from a neighboring town. She had partially paralyzed her right hand in a fall from a tree and had woken up in the middle of the night seized with an impulse to visit Lourdes. What a very pregnant woman was doing climbing a tree is unclear, but according to the story, after dunking her hand in the spring, her fingers apparently unfroze and regained normal movement. Then she returned home and gave birth to a son. There was also a man who had been blind in his right eye ever since an explosion twenty years earlier. After splashing the spring's holy water onto his eyes for several days, he too claimed he was healed. As word of these seeming miracles traveled, more and more sick and injured people started showing up in Lourdes, even though Bernadette's Virgin Mary had never suggested the spring had divine curative powers. In fact, Bernadette's own health never seemed to benefit from it.

So it was that this town in southern France, nestled into the foothills of the Pyrenees Mountains, became a place for miraculous healings. Some 6 million people a year now come to visit, some of them sick and hoping for a cure, others wanting to comfort and assist the sick, and far more just intending to pray and worship at the Catholic Church's largest and most elaborate pilgrimage site outside of Rome. The town's hilled streets host hundreds of hotels and are lined with souvenir shops selling Mary snow globes, Jesus oven mitts, and rosary beads in every shade of the rainbow.

Lourdes's ascendance onto the world stage was helped because, almost from the start of the supposed healings, the Catholic Church attempted to distinguish between genuinely interesting claims and those woven with large doses of wishful thinking. This was done with widely varying degrees of rigor, but it nonetheless gave the impression that healings at Lourdes were credible because someone was assessing and verifying them. The first such assessors were priests and church officials, who produced an unsurprisingly large number of supposed cures, often several hundred per year. A few decades later, in 1883, a medical bureau was established under the supervision of a doctor, who recruited other doctors visiting the shrine at the time of the alleged healing to help evaluate it. In 1954, another layer was added, an international committee of doctors and medical experts, not all of whom were Catholic. Over the last 160 years, roughly seven thousand people claiming a healing at Lourdes have had files opened on their case. Of these, seventy have been declared, in language adopted in the 1980s, "unexplained according to current medical knowledge" and then anointed as official miracles in a separate process undertaken by Church bishops.

Today, the doctor who oversees both the medical bureau and the thirty-two-person international medical committee is a sprightly Italian named Dr. Alessandro de Franciscis. At his office on the second floor of a long limestone building that's within view of the shrine's glittering cathedral, Dr. de Franciscis offers me thick dark coffee in a tiny plastic cup and begins telling me how miracles are born. Every year, he says, amid the hundreds of "beautiful stories" told, he and the visiting doctors of the medical bureau find that thirty to forty are worthy of serious consideration. They must meet four criteria: They need to be instantaneous (no gradual healings at Lourdes), complete, long lasting, and unexplainable by current medical science. Most cases are eventually closed and discarded, either by Dr. de Franciscis or more likely because of a person's lack of interest in fulfilling the bureau's endless requests for documentation and information. "When I start the procedure of building the clinical file, it's very painful and annoying and highly unsuccessful," Dr. de Franciscis says. "Most people are not at all interested in doing what's needed to come up with a medical resolution, to follow up on all the medical evidence. We write, we solicit, we telephone, but nothing."

Such aggravation, he says, is an inevitable consequence of an evaluation process that's far more rigorous than anything else out there. He mentions visiting a Christian Science office in Boston in the eighties when he was a graduate student at the Harvard School of Public Health. Christian Scientists believe that faith and prayer alone can cure most, if not all, diseases. "I went into one of their offices and asked them about how they validate cures. The woman that was there told me it was based on the testimony of the patient and their doctor. That's not validation."

Dr. de Franciscis is tall and thin and looks a lot like the Italian actor Roberto Benigni. Dr. de Franciscis is sitting in a stately antique brown armchair and has a habit of leaning back into it and closing his eyes when I speak. The first time this happens I panic, thinking he's nodded off during what must be an exceptionally long and rambling question. Eventually I realize that he's just concentrating closely so as to steady himself with the right response to topics he views as sticky terrain. For instance, my query about whether some percentage of Lourdes healings could be due to the spontaneous and random natural remission of disease that's known to occur from time to time.

"Yes, it could be," he says, "but this has nothing to do with me. I'm a doctor. I believe in the scientific method to get to a final diagnosis, and I do respect the point of view of someone who believes that something that is unexplainable today will be explainable tomorrow. And if someone says they do not believe in God or a supernatural force, I respect that." There is a pause. "But I think I can have the

same respect for a bishop pointing out that he believes there is an
action of God, the intercession of Mary, the mother of Jesus. Both
theories stand at the same level."

Dr. de Franciscis seems to be saying that in the absence of conclusive proof about the nature of spontaneous healings, one can go
in either of two directions, both equally valid in his view: you can
proclaim them to be benevolent acts of God or the lucky result of
some internal physiological process we don't currently understand.
He appears to be walking through the first door.

"Are you saying spontaneous remissions are acts of God?" I ask,
checking.

Dr. de Franciscis takes a sharp breath. "If the question is 'Do I
believe in miracles?,' I do believe in miracles. I truly do. Science has
not proven the nonexistence of God. I understand skeptics, and I
have respect for their views, but what I'm asking is equal respect for
those who believe this might come from the supernatural."

In addition to being in charge of rigorously evaluating healing
claims, Dr. de Franciscis is a devout Catholic. In 2014, he became a
celibatarian, which means he took a vow of celibacy. He also took vows
of poverty and obedience, which oblige him to follow Jesus's teach-
ings and the Church's doctrines. This consecrated life, as it's called,
is as close as you can get to being a priest without actually becoming one. Dr. de Franciscis is also a member of the Order of Malta, a
Catholic lay organization dating back to eleventh-century Jerusalem.
He says he doesn't see any conflict between this life of faith and his
role in determining if healings are scientifically unexplainable. "After
seven seasons in Lourdes, I have never found any conflict between my
believing in Jesus Christ and my Christian God and in my trying to
be scientific in my approach to diagnosis and cure. No conflict at all."

The urge to interpret mysterious occurrences as supernatural
lies deep in human nature. Scientists in the emerging field of neurotheology, or "spiritual neuroscience," have posited that religious faith
or belief in otherworldly forces is hardwired into our biology. The
neurotransmitter serotonin, a network of neurons in the brain's

frontal, parietal, and temporal lobes, and the so-called God gene, VMAT2, are all thought to be involved. Indeed, some things are so wonderfully unfathomable as to appear as if they could come only from a different plane of reality. But I can't muster support for the view that faith and science operate at the same level. As a system of insistent examination and revision, science is continually revealing truths about our biology and the world we inhabit, sometimes to the detriment of divine assumptions. In *On the Sacred Disease*, Hippocrates posited that epileptic fits were not a force cast upon one's soul by a particular deity. "Men regard [epilepsy's] nature and cause as divine from ignorance and wonder," he wrote. "And this notion of its divinity is kept up by their inability to comprehend it." Two millennia later scientists afforded us that comprehension. Seizures, we now know, are the result of a sudden rush of electrical activity in the brain. Maybe something similar is happening at Lourdes. As science advances our understanding of the intricacies of the human body, the terrain of inexplicable healings narrows. The vast majority of those seven thousand alleged cures occurred before 1950, when there was a lot more medical science couldn't explain or treat.

But this doesn't mean that interesting things aren't happening at Lourdes. I tell Dr. de Franciscis that I'm curious about what was then the most recent miracle, #69—an Italian woman named Danila Castelli, whom doctors had diagnosed with a rare, noncancerous tumor called a pheochromocytoma, which appears on the adrenal gland. These abnormal cells cause periodic and dramatic surges of the stress hormones adrenaline and noradrenaline, leading to all-out fight-or-flight responses. For nine years, Danila had regular such spells. She would fall victim to heart palpitations, sweating, dizziness, shortness of breath, headaches, tremors, and blood pressure as high as 280/150. After a visit to Lourdes in 1989, she was deemed completely healed. Her case grabs me because many alternative practitioners concede tumors, even benign ones, are difficult to eradicate. Also, from the videos I've seen online, Danila Castelli has something sweet and angelic about her.

Dr. de Franciscis leaves his office to retrieve the international medical committee report for Danila's case. He returns a few minutes later thumbing through it. "Danila Castelli is married to a gynecologist," he says. "He is Iranian and he helps her to go to the best doctors available at the time, in the eighties. They remove all different parts of her body—uterus, parts of her guts, adrenal glands, pancreas, but the suffering continues. The only thing she could do that helped was injections of phentolamine [a drug that lowers blood pressure]."

In total, between 1980 and 1989, the report notes, Danila Castelli had five surgeries to remove what doctors were sure was the problem. Each time, though, her symptoms returned. Then in May of 1989, she emerged from the baths at Lourdes, which are filled with water from Bernadette's hidden spring, with what the report says was "a feeling of sudden and extraordinary wellness." After that, she never had another hypertensive episode, and the things that almost always provoked them, such as sex and gynecological exams, no longer did. The report cites a gynecologist who did an exam with Mrs. Castelli several weeks after her Lourdes trip.

"Now twenty-six years later," Dr. de Franciscis says, "one can say this is lasting forever. And there is no known explanation for a cure like this."

Dr. de Franciscis arranges for me to meet with Mrs. Castelli at her home in Pavia, thirty kilometers from Milan. We discuss how I might get there, as there are only a few flights a week between Lourdes and Milan. Dr. de Franciscis decides I should leave in three days, since Mrs. Castelli, now sixty-nine, has just gotten out of the hospital for the removal of a benign hemorrhoid-like growth in her colon. She needs a few days of rest. Also, Dr. de Franciscis says, this will give me some time to explore Lourdes.

"It's not a medical institution," he explains. "Lourdes is a place of worship and pilgrimage and a place to be of service to others. You have people who set aside dollar after dollar all year to build up enough money to pay their airfare tickets to come here and volunteer with the sick for a week. You have people that sit down to listen to stories, and doctors who call the sick people by their name, not their diagnosis. I think that the miracle of Lourdes is Lourdes itself."

On my way out, he hands me a guidebook. "Go breathe into Lourdes."

I decide that one of the things I should do while here is to plunge myself into the same baths Danila entered moments before she had that sensation of "sudden and extraordinary wellness." For many people, this is a highlight of their trip here, a direct communion with the visitation that started the whole thing. A number of Lourdes miracles have occurred after a bath.

Despite their popularity, few photos exist of these baths, and I know little of what to expect. Are they communal arrangements? Does one need a bathing suit? Photos of the shrine's churches, the rock grotto, the lofty spires, the elevated walkways, and the river Gave de Pau snaking through town appear online. But the Lourdes baths remain a visual mystery. What I do know is that, sadly, they are not hot tubs. The water temperature hovers around a brisk fifty-two degrees, and even before taking up residence in Hawaii, I was never a fan of cold water. But I did want to know what Danila went through on that life-changing day, and what, if anything, about the baths supplied her with such an unusual feeling.

The entrance is a small, simple enclosure about thirty feet from the edge of the river. When I get there, several dozen people are waiting in silence on wooden benches, men on the left, women on the right. Many of the older women twirl rosary beads in their fingers. People in wheelchairs and movable beds, each flanked by teams of white- or blue-clad helpers, have their own line. When my turn comes, I am ushered behind a royal-blue curtain into a room the size of a freight elevator in which no fewer than eight women are dressed in white and ready to assist. Each holds up a piece of cloth the same shade of royal blue, making it look as if we were all swimming together on a Hawaiian beach.

"Take off everything but keep your bra in your right hand," one of the white-clad ladies tells me while holding up the blue tenting. I set down the bag into which I've foolishly stuffed my bathing suit and do as I'm told. With all my clothes off and my bra dangling from my fingers, the blue barrier collapses around me like a robe. Someone swipes my bra and nudges me through another curtain, revealing a deep rectangular bathtub cut into the gray marble floor. Just looking at it makes me feel cold. Since I'm tall, I figure maybe I'll just wade in up to my thighs or so. That should do it.

One of the volunteers, a middle-aged Irish woman with a soft voice, wraps her arm around me and hums in my ear. "Let it happen," she says. Pointing to a small porcelain statue of Mary on the wall, she urges, "Go ahead and give your intention to the Virgin Mary and then make a sign of the cross with your right hand." How long has it been since I've done that? I fumble around and get the sequence of taps wrong—the only person, I have to imagine, who comes to a Catholic shrine and does such a thing. Quietly, I apologize to all the nuns who steered me through thirteen years of Catholic school. "Don't be nervous," my Irish friend says. "Everything is going to be as it's supposed to be."

The dry blue cloth around me is peeled off and replaced with a cold, wet white one, and I am led by two women, one clasping each hand, into the bath. "Walk to the front and don't kneel. Sit down," one says, dashing my plans for partial immersion. I let my legs go out from under me until the water rises past my belly button, filling my lungs with a burst of air too big to handle. I gasp and struggle to exhale. The water rises higher until my butt hits the hard bottom. I spend a few seconds like this, trying to breathe, then, before hypothermia sets in, I am guided from the water. My friend is ready with the blue cloth. "Takes your breath away," I say, still struggling to inhale air properly.

She nods knowingly. "It certainly does."

A few moments later, I'm back outside in the afternoon sun, noticing its glare ripple along the river. It's hard to say how long I was in there—time seemed to stand still. But I'm guessing it took no more

than ten minutes. Feeling light and airy and dry, I sit along the stone
wall next to the river, watching people disappear into and emerge
from behind the curtains. Nobody looks especially euphoric or collapses onto the ground after his or her bath, but people's expressions have a solemn vulnerability. Even if they're not sick, everybody
comes here for something, and contemplating those private moments
of need and surrender, I feel a stirring of tremendous awe.

The thirty-minute train ride from Milan to Pavia slices through
alternating industrial and agrarian terrain, as if no one decided what
kind of economy the area is supposed to have. Although considered
a suburb of Milan, Pavia thrives mostly as an academic center, with
half a dozen universities, including one of Europe's oldest. That ven-
erable institution, the University of Pavia and its medical school, is
what first brought Danila's Iranian husband to the town where she
grew up.

Danila and her daughter Valeria meet me at the train station roundabout in their small Fiat. Valeria has come along to translate since
Danila speaks little English and my Italian stops at the end of a restaurant menu. We spend the ride talking about my visit to Lourdes and
Danila's brief stay in the hospital, which she reports went smoothly.
"I am feeling pretty good," she says. Her outfit seems to confirm it: a
formfitting black-and-white dress and black high-heeled shoes.

Only when we get out of the car do I see how years of suffering and endless medical procedures have taken their toll on Danila's
body. Her silhouette appears as one straight, preadolescent line, and
she looks no heavier than the bag I've been lugging around with
me for the past week. I feel a strong urge to make sure she doesn't
stumble on the walk to the house, seeing her frailty and recalling
how many body parts she's had removed by surgeons trying to help
her. Her uterus, fallopian tubes, ovaries, and left adrenal gland are all
gone. As are parts of her pancreas, rectum, and bladder. The hardest
of all, Danila will tell me, was the loss of a portion of her stomach,

because of how difficult it makes eating. Her meals must be consumed in numerous child-size snacks throughout the day, making putting on more weight nearly impossible.

The house in which Danila and her husband, Farhad,* raised their four children is divided into two parts, linked by a large concrete courtyard. On one side live Danila and her husband, on the other Valeria, her husband, and their toddler son. Valeria is the youngest child and, like her dad, a doctor. Her field is dermatology. We sit down to talk in the kitchen on Danila's side of the house, where a lingering smell of old cooking oil with low notes of onion and garlic suggests many busy years have unfolded here.

"I dreamed of going to Lourdes as a little girl," Danila says. "I was always very faithful and wanted to become a nun, but then I fell in love with my husband and my mind changed." With her small arms resting on the table, she grins.

"I went first in 1983 to Lourdes when I was sick, and then another time a few years later, both times with a priest and the church community here. Then in 1989, when I felt like I was dying, I went with my husband."

I'm certain Dr. de Franciscis hadn't mentioned Danila's 1989 healing was on her *third* visit to Lourdes. In the news conference announcing the miracle, he noted she had "desired all her life to come to Lourdes." I suppose he hadn't explicitly said it was her first visit, but that was certainly the impression. The medical report also made no mention of previous visits.

"So there were two visits before the healing?"

"Right," Danila says. "Those two other times I went with the church."

The first two times she visited, Danila says, she wanted her husband, who at the time wasn't Catholic, to come with her, but he refused. After nine years watching his wife suffer, though, Farhad began to reconsider. He had made an appointment with endocri-

---

*Danila's husband's name has been changed.

nologists at the Mayo Clinic in Minnesota in the hopes they could find answers that had eluded the Italian doctors, but he was worried about his wife making the long journey. Instead, he suggested they drive to Lourdes.

Except for his presence, everything about Danila's visit was similar to her two previous ones. She attended mass, prayed in the churches, and lit candles in the Torchlight Procession. Her dip into the healing baths wasn't new either. She had done that both times before, and on her second visit she had made the same plea. As she stood enshrouded in the cold, wet sheet, facing the statue of Mary, Danila didn't ask to be healed of her dreadful episodes of high blood pressure. She wished for her husband to find a sense of peace before her death, which she felt was imminent.

"My father was so angry because he is a doctor and he thought his colleagues didn't treat his wife right," Valeria says. "Her illness was like a pinball game. The tumors were here, and then there, and they would cut, cut, cut. He was ready to sue the doctors for malpractice. My mom didn't want him to do that. She said what's done is done."

When Danila emerged from the baths on that third visit, Farhad saw that something was different. An unusual happiness was in his wife's face, a joy he hadn't seen for years. He embraced her and whispered in her ear, "I know that everything now has passed. I know that everything is behind us." He had no idea why he knew this. As a doctor, it made no sense for someone's long-term health problem to dissolve so suddenly. He could not even begin to explain such a thing. Also perplexing was that he seemed to know it with such certainty. He told Danila he was prepared to forgive the doctors for the unnecessary surgeries and drop the malpractice lawsuit he had been preparing.

Danila was elated, but felt her husband was crazy to presume she had been healed. Although she had a powerful feeling as she walked out of the baths, she didn't imagine it had anything to do with her disease. "An indescribable joy," she says, trying to describe it. "A great freedom." Looking unsatisfied, she tries again. "A happiness that is not human." But this also doesn't seem to do it justice. She apologizes for not having better words to capture the feeling, which she says occurred only one other time in her life, when she was under anesthesia from surgery in the early eighties. She awoke in the ICU after having had such a beautiful experience while asleep that she was angry to be back in her old life and inside the same troubled body, racked with postoperative pain. She longed to go back. But just as that experience hadn't led to a disappearance of her health problems, Danila saw no reason to think this one had either.

Even when, later that night, the couple had sex for the first time in many years without triggering a hypertensive episode, Danila still didn't believe her affliction was gone. She awoke the next morning and asked her husband to give her an injection of phentolamine, which he had been doing multiple times a day to keep his wife's blood pressure spikes at bay. The drug, which is sold under the brand name Regitine, causes blood vessels to expand, thereby increasing blood flow. Farhad assured Danila she didn't need it, but she insisted. When he gave her the dose, it made her feel nauseous, which hadn't ever before happened. It was the last time she ever took it.

Valeria, then fourteen, says she remembers her parents being unusually happy when they got home. She also noticed that her mother was no longer getting the injections. "My father told us it was a new medication," she says. "They didn't know what was happening, so they didn't tell us anything. Sometimes when you're dying, you have a little period where you feel really good, and then you die. So, they thought maybe it was like that."

Valeria ferries three small espresso cups and a bowl of sugar to the table. Danila looks at me and says, "When you live almost ten years in hell, the first day you are in heaven, you don't ask."

The room is quiet for a moment, then Danila asks me what I think about her healing: What do I think happened? I'm not expecting this question and I blurt out something incoherent about my efforts to understand mind-body healing. I say it seems possible that the emotions she felt having her husband on that long-awaited trip could have contributed to her healing. But the reality is, I don't know of any mechanism by which joy, rapture, or any other state of mind, no matter how absorbing, could instantly eradicate abnormal cells on one's adrenal gland or anywhere else. Was it some inexplicable natural healing that could just as well have happened while Danila was home making dinner or out taking a walk, and that it took place at a sacred healing site a cruel and deeply distracting coincidence? Maybe, but why would cells that cause surges of adrenaline and noradrenaline suddenly burn themselves out after nine years of stubborn interference? That didn't make much sense either.

I know Danila attributes her cure to God's benevolent intervention, a biblical water-into-wine and blind-man-seeing marvel. She is pious, and in the absence of another explanation, this, to her, is a natural, as well as a deeply gratifying and inspiring, conclusion. For me, though, it's perplexing to think about God selectively sending down cures to a small town in France. How many other sick people were at Lourdes that day and went back home with the same physical ill health? And why did God heal Danila on her third visit, not on one of the other two?

"We have to ask God," Danila says, shrugging elaborately. "Maybe it just wasn't my time before that [visit]." She says she thinks it's possible that the relief and happiness she felt being at Lourdes with her husband may have contributed to what happened, but she hardly considers it sufficient. "I think you need to have faith. Man alone doesn't have this strength."

Danila says she has done many events with churches and other groups since the healing. Often, people want to speak with her, touch her, or give her messages to pass along to God. Some think that she can see Mary. Danila has to tell them she isn't the living act of God they imagine her to be. "I have no power to heal people, and I have no direct link to God. I'm just someone who's deeply grateful to still be alive." She will talk at length with these people, often holding their hands. Others, though, are resentful. "They think, 'Why doesn't God talk to me or heal me?' A lot of people ask why you and not me. There is no answer."

On the train back to Milan, I leaf through Danila's ten-page medical report again. I plod through dense sections strewn with perplexing terms such as *adrenalectomy*, *zona glomerulosa*, and *symphysis pubis*. But a few sentences jump out at me. "Apart from the left adrenal lesion," the report notes, "the various histological examinations made following surgical procedures have not demonstrated the presence of other neoplasms." This means that the microscopic examinations of tissues taken from Danila's body after the surgeries showed none of the abnormal growth you would expect to see with pheochromocytoma or paraganglioma, which is the name given when these tumors are found in areas outside the adrenal glands. None, that is, except for one small tumor on Danila's left adrenal gland, which was removed in 1986 along with the rest of the gland. But strangely, this removal didn't result in the improvements her doctors expected. In fact, there was virtually no improvement. If those abnormal adrenal gland cells had really been the problem, taking them out should have had a measurable impact.

It seems possible then that Danila never had pheochromocytoma in the first place, and that the one tumor she did have wasn't the problem. Confused, I email Dr. Fausto Santeusanio, the endocrinologist and Lourdes medical-committee member who authored the report, to ask him where, if anywhere, Danila's tumors resided. His answer doesn't offer much new information. Her pheochromocytoma, he says, was supposed to be in the left adrenal gland, but apparently wasn't, and that Danila regularly went into a hypertensive crisis during gynecological exams suggested a tumor might have been somewhere in the pelvic region. "As you can see," Dr. Santeusanio wrote to me, "the case is very complex and I'm unable to give more clear and convincing answers to your questions."

It's strange to think that a case considered for more than two decades before being declared an official miracle would still have big unanswered questions, or that the diagnosis would be shaky. It becomes even more precarious a few months later after a conversation I have with Dr. William Young, the head of endocrinology at the Mayo Clinic, where Danila almost went for treatment. "Most patients whose doctors diagnose them with pheochromocytoma don't have it," he tells me. "This is an incredibly rare tumor." Out of one hundred people who appear to have it, Dr. Young says that maybe only one or two actually do. They could have any number of things, but by far the most common cause of dramatic hypertensive episodes is a problem that has nothing to do with tumor growth or any other medically detectable disease. "I'll see patients at Mayo Clinic who have had one of their adrenals out to treat pheochromocytoma. They still have symptoms and they come to see me to find their tumors, and I have to tell them, 'Look, you have panic disorder; it's a good thing you have two adrenal glands and you only need one.'"

As any of the estimated 12 million American adults who will suffer from panic attacks at some point in their lives can tell you, getting an episode feels as if you're having a heart attack, can't breathe, and are about to die. But nothing is wrong with the heart, and you will live to see another day. Panic attacks are the body's stress response rearing up for seemingly no reason, when there's no need for the associated physiological changes such as increased blood flow and heart rate, rapid breathing, and sweating. Scientists don't quite understand why these episodes occur, but the prevailing theory is that they are a reaction, either conscious or unconscious, to subtle changes in the body, such as elevated blood levels of carbon dioxide or a faster heartbeat. Yet these bodily fluctuations aren't really the cause—plenty of people experience them without issue. It's thought that panic attacks occur when a suffocation alarm in the brain is triggered by elevated carbon dioxide or a wave of anxiety comes from a racing heart. These mental and emotional reactions fire the stress response and generate more unusual bodily sensations, making for a vicious cycle. It's not unlike the way relatively minor signals from our tissues can trigger the brain and induce chronic pain.

What makes someone hyperreactive to certain physiological alterations is unclear, but the theories look similar to those for chronic pain as well: the presence of psychological and emotional stress, and a genetic predisposition for a particular type of brain wiring or imbalance of neurotransmitters. Cognitive behavioral therapy can, as it does with chronic pain, help reduce panic attacks by teaching people how not to let their thoughts and fears escalate when they feel the bodily sensations of a panic attack coming on.

"Most patients have no idea that this is going on," Dr. Young says. "Everyone tells me they're supercalm. But it's things in their life that they're not recognizing and not dealing with, and it's a matter of visiting a psychiatrist and working through those issues and also getting put on certain medications that can really help."

A related and lesser-known condition that doctors have dubbed pseudo-pheochromocytoma was first named in 1956. Popularized in the eighties by a Canadian endocrinologist, it differs from panic attacks in that blood pressure soars to an even higher level. Dr. Samuel Mann, a hypertension specialist at New York-Presbyterian Hospital and professor at Weill Cornell Medical College, believes pseudo-pheochromocytoma is driven not by current stress and anxiety, but by repressed emotions related to past trauma. The mechanism is unknown, and his views are based on observational work with patients and what he says is a high success rate of treating the condition with antidepressants.

When I send him a copy of the international medical committee report Dr. de Franciscis gave me, Dr. Mann calls me back a few days later to tell me he's quite certain Danila didn't have pheochromocytoma. This disease, he says, always produces off-the-chart levels of stress hormone by-products, on the order of a four- or fivefold increase, and Danila's blood tests showed only "moderately elevated" levels. Moreover, the radioactive tracers, or MIBG scans, that were used to identify Danila's areas of abnormal cell growth are now known to produce a lot of false positives, which wasn't widely understood at the time she was undergoing her testing. Both Dr. Young and Dr. Mann say they use MIBG scans only when blood tests reveal markedly elevated hormone levels and when imaging scans don't pick up a tumor on the adrenal glands, suggesting it might reside elsewhere. "I would never order an MIBG if the catecholamine levels are normal or near normal," Dr. Mann says.

He thinks it's likely Danila had pseudo-pheochromocytoma, and his first guess, given that sex and gynecological exams were among her triggers, is that she may have experienced rape or sexual abuse, as an estimated 13 to 34 percent of Italian women have. "That's pure speculation," he notes. Regardless of the cause, he says he would have treated her with an antidepressant and, if necessary, an antianxiety benzodiazepine such as Xanax. Psychotherapy would have been an option if she was open to exploring a possible link to prior trauma. "This is hard stuff to get to, and sometimes I don't even know if we're meant to get to it."

Asked whether Danila's Lourdes visit—her dunk in the healing waters, the resulting joy, and her husband's sudden and vigorous decision to forgive all the doctors who had failed her—somehow "got to it," he says perhaps. "Our belief in miracles has a fantastic placebo effect."

I email Dr. de Franciscis about this and am surprised to hear his response: the pheochromocytoma diagnosis, which was highlighted at his press conference and mentioned as the probable diagnosis in Dr. Santeusanio's report, is not certain. He writes, "The diagnosis of the unexplained cure of Mrs. Castelli was, and remains, that of 'severe and recurrent life-threatening crisis of arterial high blood pressure.'" He says he would have liked to examine tissue samples taken during some of Danila's surgeries, but they were no longer available.

As to what may have caused such life-threatening high blood pressure, he doesn't know. He tells me *he* is surprised to hear *me* raise the possibility of psychological causes and says that Danila never underwent psychological or psychiatric evaluations. "It was never needed because of the impressive organic syndrome she presented for almost ten years," he writes. (By "organic syndrome," he means physical or biochemical.)

I want to ask Danila whether she ever considered the possibility that the intense physical symptoms she experienced for nearly a decade were psychologically rooted. Individuals rarely assume this to be the case. When our body goes haywire with measurable problems such as high blood pressure, we don't believe it's the brain wreaking havoc. Yet panic disorders teach us this can happen. I wonder if after so many useless surgeries, Danila might have started to speculate, even if Dr. de Franciscis never did.

But when I get in touch with Valeria via email, she gives me the sad news that her mother passed away a month earlier, in October 2016. She died almost a year after I met with her—of what Valeria won't say. She asserts that she is unable to answer any of my follow-up questions and refers me back to Dr. de Franciscis.

It would seem that a full accounting of whatever made Danila Castelli suffer nine years of high blood pressure spells so severe that she thought she was dying is going to remain a mystery. We don't know whether psychic pain from a past trauma or another factor might have contributed to her condition. What's clear, though, is that what disappeared that day in May of 1989 wasn't a tumor. More likely, it was psychological distress that for years had triggered physical symptoms. Rather than prayer or bliss mysteriously evaporating "organic" maladies at Lourdes, it is probably this kind of hard-to-bring-about emotional cleansing that explains others of its supposed miracles. Like placebos, states of "indescribable joy" shift the mind, and the mind shifts the brain and body in turn, though how such a transformation might work neurologically remains a mystery.

But in my view, this doesn't make these events any less exceptional. People struggle for years to exorcise their demons. Danila's problems, whatever they were, vanished almost instantaneously, freeing her from the prison of long-term suffering. It is, in all its wonder, still a miracle.

All in My Head

The German psychosomatics

Just a simple utterance of the word can be toxic. Most doctors will do everything they can to careen around it, lest they appear to be telling patients, "There's nothing wrong with you. You're making a fuss over nothing." Or worse yet: "It's all in your head." Instead of declaring an illness to be *psychosomatic*, doctors will reach for any number of other descriptors to communicate the presence of symptoms they can't fully explain medically—maladies that every conceivable test and imaging scan has failed to link conclusively to a bodily disorder. One doctor told me he summons the word *stress* because "almost everyone can agree that *stress-related* is something that applies to them, at least somewhat." *Stress* evokes socially acceptable levels of busyness and the daily dramas of life, not the sensitive chaos of inner turmoil. It also has the distinct advantage of not containing the prefix *psycho*.

Patients might also hear they have a *functional illness*, which sounds sufficiently medical but offers no actual information. A *functional* problem merely refers to the idea that some system of the body, though undamaged, is not functioning correctly, which comes as no surprise to people who find themselves in a doctor's office. What patients really want to know is why their symptoms are happening and how they can be fixed—queries doctors sometimes have no clear answers for. A far simpler solution is to avoid the subject altogether and send patients off for more tests, supply them with a new prescription, or refer them to another specialist. Anything to avoid saying that someone's suffering might be arising to a significant degree from the thoughts, feelings, beliefs, and memories that have wired themselves into his or her brain. To many patients, such a diagnosis sounds as if they are being called crazy.

The only place in American medicine where *psychosomatic* is allowed to roam free is in academic circles—in the gatherings of the American Psychosomatic Society and its publication, *Psychosomatic Medicine*. There, doctors and psychiatrists view medicine the way the German physician Johann Christian August Heinroth, who first coined the term in 1825, had in mind. Merging *psychik*, the Greek word for "mind" or "soul," with *soma*, the Greek for "body," Heinroth wrote, "The person is more than just the mere body as well as more than the mere soul: it is the whole human being."

Yet despite a sensible mission of integrating the "mind, brain, body and social context in medicine," the American Psychosomatic Society has not had an easy time of it. Over the years, members have periodically floated alternate names, none of which ever seemed quite right. There was the American Society for Behavioral Medicine, but this felt deaf to the emotions and thoughts that drive behaviors. The American Stress Disorder Society implied that people needed to quit being such overachieving do-it-alls. The American Mind-Body Medicine Society sounded like something Deepak Chopra would be the president of. Only the American Society for Biopsychosocial Medicine managed to sum it all up, but what a mouthful. So, in the end, *psychosomatic* stuck. For the most part, it was safely tucked away at conferences patients didn't attend and in journals they didn't read.

But this isn't the case in Germany, where *psychosomatic medicine* is a subspecialty of internal medicine and dares to hang its shingle in broad, shameless daylight. Here in Tübingen, at the University Medical Hospital, the psychosomatic department shares space inside the hospital's busiest hub, a few floors up from the endocrinology, pulmonology, and gastroenterology departments. Down a nearby hallway

is the ICU. I'm here to understand why the Germans are so unperturbed by this word and what exactly they consider psychosomatic.

Located two to three hours west of Munich by very timely trains,
Tübingen is a university town in southern Germany that's capti-
vating in all the old-world ways you would expect. It is fronted by
mountains and intersected by a river, and its ancient, medieval buildings are interspersed with colorful, triangle-roofed Bavarian houses
that I can't look at without thinking of Hansel and Gretel. On the
week in late June that I visit, a spell of what I'm told is unusually
steamy weather seems to have put no dent in the town's normal buzz
of activity. Everyone is zipping around town on their bicycles (*ohne*
helmets), navigating designated bike lanes in dense, speedy clusters.
I see so few people in their cars that I wonder how someone who
wasn't skilled at riding a bicycle could live in this town, at least during the summer months when Germans flock to the outdoors.

Although Tübingen has one of the country's largest and most
established university psychosomatic departments (and one of Germany's largest hospitals, with sixteen hundred beds), I could theoretically have gone to several other university-run hospitals in Germany
with similar departments—Heidelberg, Munich—or to one of the
many separate psychosomatic hospitals in the German countryside,
often beside a lake.

The view from Tübingen's psychosomatic medicine department,
up on the eighth floor and overlooking town from a hill, isn't bad
either. "All the rooms have these views," Dr. Florian Junne, the department's deputy director, says as he motions toward the valley and
sweeps his hand across the triangled tops of buildings. "We wanted
the patients to be able to have this to look out on." Here is where
psychosomatic patients come, thirty at a time, as if attending summer camp. They arrive with their clothes, their toothbrush, and other
personal effects for a stay of six to eight weeks, going home only on
weekends. To land at a German psychosomatic medicine department
is to make getting well your sole purpose. People with jobs take paid
leaves—in Germany such things are universally possible. Those who

are in school take semesters off. And nobody goes into bankruptcy. Like most other medical treatments here, the lengthy stays are covered by Germany's public health insurance system.

The affable Dr. Junne is my tour guide for what is warmly, but I think unfortunately, called the ward. To me, a ward is where Nurse Ratched does her rounds, and that, Dr. Junne assures me, is not the agenda here. Patients come here voluntarily and spend their days undergoing a schedule of pleasing and self-empowering, if at times challenging, mind-body treatments. These typically run from eight in the morning until around four or five in the afternoon. When I ask Dr. Junne to tell me about this word *psychosomatic*, he says that he and his colleagues define as it as "the interplay of psychological factors with bodily sensations and symptoms." This does not imply, he says, that patients are making themselves sick on purpose or acting worse off than they are. "All of this is one hundred percent real for these patients."

Generally on the ward, there are four categories. The biggest is chronic pain, which few doctors in the US would dare taint with the *psychosomatic* label, but nonetheless fits based on everything we know about the brain's role in pain's manifestations. Also here are patients whose cancer, heart disease, or other life-altering medical problems have caused them high levels of anxiety, stress, or depression. They come, as the Harvard psychiatrist Arthur Kleinman might say, to treat the illness arising from their disease. The last two categories are eating disorders and so-called conversion disorders, which are fascinating conditions that can look like multiple sclerosis, epilepsy, or some other neurological disease, but aren't. Someone might have twitching, drooping, or otherwise uncooperative muscles or might fall to the ground in what looks like a seizure. Sometimes people even lose their ability to walk or become partially blind or deaf, although nothing is wrong with their legs, eyes, or ears. The thinking on this, much like Dr. Samuel Mann's theory about pseudo-pheochromocytoma, has always been a bit Freudian: that the brain unconsciously "converts" some emotional stress or trauma into neurological symptoms. No one is sure scientifically what the mechanism is or how much biochemical or genetic predispositions in the brain might factor in. Studies show that such conversion symptoms—or functional neurological symptoms, as they're now called—aren't as rare as you might think. Between 1 and 9 percent of people showing up at neurology offices have such medically unexplainable symptoms.

Dr. Junne walks around two floors of blindingly white and clean hallways opening doors. We peer inside small, cozy areas where patients meet with a physician or psychologist twice a week for individual psychotherapy sessions and bigger ones where they convene in circles for three weekly group cognitive behavioral therapy classes and, for chronic pain patients, pain education classes. "We talk, for example, about central nervous system sensitization and how pain gets amplified by different factors, and why in the case of chronic pain, the intensity of the symptom is not a very accurate measure of physical damage in their body," he says, echoing an increasingly familiar refrain.

"Here, we try to enable patients to use their mind to slow their heart rate and change measurements of their muscle tension. The aim is to improve their stress-coping skills and sense of self-efficacy towards their body," Dr. Junne says, peering into the room with the biofeedback machine. It is next to a space with a turquoise massage table where patients can get different types of massage. Down the hall is the meditation room, where progressive muscle relaxation is also taught. In this therapy, patients practice systematically tensing and then relaxing individual muscles all over their body. Next door, in the physio room, which is most heavily used by chronic pain patients, people work on what Dr. Junne calls "a sense of empowerment."

"What we want is for patients to be active participants in this change process, because a lot of them really haven't had that for so long. Because of massive pain symptoms, they've gotten used to not doing things. After months and years of suffering, they have convictions such as 'I can't do Nordic walking' or 'There's no way I can swim; I just want to get a massage.' So we try and explain how this kind of avoidance behavior makes things worse, and how moving their body is not going to harm it," he says, skirting around a stockpile of yoga mats, jump ropes, and big exercise balls. If stretching and Pilates aren't your kind of empowerment, there's also swimming at the hospital's pool, Nordic walking outside with poles, and qigong classes.

Or self-confidence can come for people through art therapy or down in the subterranean music therapy room. Next to the freight elevators and far out of earshot of others, music therapist Jan Harnisch tells me how patients naturally gravitate toward the drums, even though the room is stocked with every conceivable instrument one doesn't need talent or lessons to play—maracas, cymbals, kalimbas, rhythm sticks, castanets, wooden blocks, tambourines. "A lot of the chronic pain patients come in and initially they play with no rhythm or modulation. It's like *bah, bah, bah.* They're used to being so protective that there's no flexibility in their body." About half of them manage to loosen up. "They figure out how to express their emotions, and they end up achieving something by playing with other people. They create something really interesting that they didn't think they could do."

By the time Dr. Junne finishes the tour, I realize the only major psychological or mind-body therapies not represented here are acupuncture and hypnosis. He and his staff run a true kitchen-sink program for altering the brain activity that's given rise to, or contributed to, different symptoms. There would seem to be something for everybody, though patients are afforded only a degree of choice in their daily activities. Part of the point is to build confidence by making people comfortable with things they thought they couldn't do. Which begs the question of whether engaging in all these mind-changing, stress-management therapies every day for six or eight weeks helps people more than, say, embarking on just a meditation or exercise regime or foraying into talk or cognitive behavioral therapy.

Dr. Stephan Zipfel's answer is ambiguous. A tall, trim man with spiky gray hair, Zipfel is the longtime head of the department. He's also the dean of medical education for the university's medical school, a post he's held for almost as long. Many of the patients who come here, he says, leaning back in his immaculately sparse office, have been ill for a long time, five to seven years on average, which can make their conditions hard to treat. Just because a certain pattern of neural firings in the brain has created or exacerbated a problem doesn't mean that treating the mind will necessarily erase it. "If someone with a conversion disorder comes in, in a wheelchair, they will often leave in a wheelchair. We have very intensive physio on our ward to strengthen their leg muscles, but it's not that easy to get them walking again." Patients who come in earlier in the progression of their condition, Zipfel says, have better chances at significant improvements.

Compared to conversion disorders, chronic pain can be easier to treat, though patients aren't told to expect their pain to go away. "We try to tackle levels of depression and anxiety first and to improve quality of life. Then we address their handling of the pain." On average, people come in with pain that's a seven to nine out of ten, and they leave with levels between two and seven. "It's more modulated and not so fixed."

Zipfel says he would like to get even more chronic pain patients treated here, particularly in the smaller day ward, where patients come and do treatments from nine to four for about four weeks, and he's trying to get the insurance systems to approve it. "I'm convinced there are a lot of chronic pain patients out there that would be better treated with psychological, mind-body, and quality-of-life treatments instead of just drugs. We're not saying it's only the mind and not the body, or that stress is the only reason. It's really the interaction that's important for the chronification of someone's illness."

To find patients that might benefit from psychosomatic treatment, Dr. Zipfel and his team meet regularly with the hospital's pain management, anesthesiology, neurology, oncology, and cardiology departments to jointly identify people with high levels of emotional stress, a history of trauma, or significant discrepancies between their symptoms and objective findings from imaging and other tests. "We sit down with patients to see how motivated they are to do this work and come up with a personalized plan," he says.

I had come to Germany thinking that perhaps the Germans, in their boldly pragmatic ways, had summoned a courageous acceptance of the mind's role in health, but it turns out that here too *psychosomatic* gets shudders from patients. "It has better connotations, yes," Dr. Zipfel says. "But people here still worry about what it means. Many patients have heard from doctors that 'nothing is wrong,' so we try to be very clear that it's not imagination. If a patient comes to us, they are not being stamped with psychiatry. They are coming to a medical hospital."

Patients with chronic pain will hear about research done with fMRI brain scans that, at long last, roots their symptoms to some area of the body. Those with back pain, for example, will learn that their brain looks different in several crucial ways from those who are pain-free. Those with irritable bowel syndrome will see how, when they experience gut pain, their brains look similar to those of people who have ulcerative colitis or Crohn's disease, conditions with identifiable tissue damage. Those with conversion disorders will learn of greater neural connectivity between the motor and the emotional processing areas of the brain. Bringing forth discussion of the *brain* apparently gives patients more ease than expositions on the *mind.*

For confidentiality reasons, I don't get to mingle with the patients, but one is interested in talking with me in the hopes that her story might help others struggling with the same problems. Anne is fifty-eight and has been on the ward for five weeks. While her fellow chronic pain patients do afternoon qigong class, she sits across from me in one of the group therapy rooms and tells me that for fourteen years she has had facial pain that was diagnosed as trigeminal neuralgia, an irritation of the trigeminal nerve of the face.

It started in 2003 as a strange, almost electrical sensation tearing across the right side of her face, from her ear to her nose and down to her mouth. Anne hoped it would go away, but when it didn't, she wondered if she had a toothache. When that was ruled out, she went to her doctor, who gave her pain medications. But these didn't seem to help; the pain got worse. After about a year, it was so bad that she would sometimes have trouble eating and talking. It was as if someone had taken a vise and tightened it around her face. She quit her job taking care of an older woman and was barely leaving the house. A neurologist eventually diagnosed trigeminal neuralgia in 2005 and recommended surgery to create some space between the root of her trigeminal nerve, near the ear, and the blood vessel that was probably impinging on it. After this microvascular decompression surgery, for which a big patch was shaved on the side of Anne's head, she felt about 50 percent better, which was a big relief. But it also suggested that the blood vessel was only part of the problem. It hadn't entirely dissolved her pain, as she and her doctors had hoped. But at least she could now talk and eat normally again, without giving all her attention to the pain. Once in a while, she could even forget about it. She started working again and focused on trying to get her life back.

Then her uncle, to whom she was extremely close, died unexpectedly. He and Anne were the same age and had always felt like twins. As she was mourning his death, she felt the pain in her face roar back. Now it wasn't just on the right side; it had spread to her neck and shoulders. Once again eating and talking became a challenge, and Anne was devastated. "I wanted to jump out of a window," she tells me, her words translated from German by Dr. Caroline Rometsch, one of the doctors on the ward.

Anne sought out more neurologists, this time here at Tübingen, two hours from her home. They recommended another surgery, which she was considering. Then Dr. Rometsch learned of Anne's case and suspected that because Anne's pain had persisted for so long, had failed to respond to surgery the way you would expect it to, had migrated to other parts of her body, and had intensified upon her uncle's death, psychological factors might be driving it. After meeting with Anne, the doctor concluded that a cramped trigeminal nerve definitely wasn't the whole story and that Anne probably shouldn't undergo more surgery.

When I ask Anne about these psychological factors, she replies that she had "a lot of trauma and violence" in her childhood.

I ask what kind of trauma.

"A lot of very difficult things." She averts her eyes and looks uncomfortable.

Anne is sitting tightly in her chair, knees pressed close. She looks neatly put together in a khaki-green dress and big beady necklace. Black reading glasses rest atop her straight, shoulder-length hair. Nothing about her connotes sickness, which only makes her problems more challenging.

Instead of pressing her about the trauma, I ask what it's been like to live on the psychosomatic ward for five weeks. This instantly perks her up. Her pain, she tells me, was a seven when she arrived and now is a three or four. She also feels as if she's becoming a different person. "I definitely have better self-esteem, better self-knowledge, and better strategies for dealing with my pain." She adds that she's found the individual therapy sessions to be the most beneficial.

Although Anne doesn't want to talk about what happened to her, she's given Dr. Rometsch permission to discuss it, so after Anne leaves, I hear some of the details that have emerged from her psychotherapy sessions. Dr. Rometsch says that Anne, like a disturbingly large number of children, was physically abused by someone who was supposed to be taking care of her. In Anne's case, it was her step-grandfather, who was raising her with his wife, Anne's maternal grandmother. Anne never knew her mother; she had left Germany when Anne was still a baby. When her step-grandfather became angry, he would insult and hit young Anne, one time even breaking her arm. This abuse caused her great distress. The biggest blow came when she was thirteen, as her step-grandfather was dying. In a quick declaration that she didn't care to elaborate on, her grandmother told her that the man who lay dying was actually her father. Thirteen years earlier, he had raped Anne's mother and, in an attempt to cover up the sin, had sent her away to the US after Anne was born. This was the real reason she had grown up without a mother or a father.

It also meant that her uncle was really her half brother. In another dysfunctional twist, he had been born to Anne's grandmother and step-grandfather five days after Anne was born. Nine months previously had been a busy time for this abusive man.

Still feeling profoundly ashamed of her twisted past, Anne almost never speaks of it. The only people in her life who know are her husband and, as of quite recently, the Tübingen psychosomatic staff. Wanting to protect her three children, now in their thirties, from something so horrible, she never told them. She has decided that she now wants them to know the truth.

As I wind my way down the twisty, tree-lined footpaths leading away from the hospital and back into town, I think about why the notion of a psychosomatic condition is so noxious. Why do we cringe in guilt and self-doubt at the suggestion? One doctor in Heidelberg told me of patients who react as if a psychosomatic diagnosis were the worst possible option: "For their first reaction, some patients even say, 'I would have preferred that it was cancer so at least I could know what was wrong.'" And this is in Germany. You have to wonder, why is it more legitimate for our body to create cancer cells or block our arteries than for our brains to generate a conversion disorder or to react to bodily signals in a way that intensifies or perpetuates pain?

Part of the answer seems to lie in that many doctors are still uncomfortable with talking to patients about the mind and brain's role in creating or exacerbating bodily symptoms. Even if the word *psychosomatic* doesn't come up, patients often still get the message that their doctor is no longer interested in helping them or able to. Before coming to Tübingen, I talked to Dr. W. Curt LaFrance Jr., a neurologist and psychiatrist at Brown University, who treats patients with conversion disorders at Rhode Island Hospital. He related watching videos in which doctors attempt to deliver a conversion disorder diagnosis to patients who suspect they have epilepsy or multiple sclerosis. "Some of the doctors are just stumbling and fumbling over their words, and they don't want to scare the patient with words like *psychological* or *psychogenic.* Oh, it's just painful." LaFrance sympathizes, as it took him many years to figure out how to best explain the problem without being either confusing or dismissive. A conversion disorder is a "neurological manifestation of an underlying psychological conflict or stressor," he now tells patients and their families.

"Rather than spending ten minutes saying, 'It's the brain or not the brain. It's in your head, but not all in your head,' I treat it as no different than any other diagnosis that we give, whether depression or migraines or stroke or panic attacks." LaFrance then talks to patients about their lives and upbringings and tells their stories back to them, highlighting possible predisposing, precipitating, or perpetuating factors as pieces of a puzzle—say, a history of abuse, a car wreck, a head injury, the loss of a job. "The majority of the time, patients and their families appreciate this clear and straightforward explanation. When they feel heard and understood, people are ready to engage in the next step, which is treatment."

Another reason for the resistance, I think, is the inherent challenge in trying to figure out when exactly the mind and the brain are driving symptoms and when they're not. We can't peer inside our heads to see how past traumas or negative beliefs have wired their way into unrelated brain regions, the way you would examine cancer cells from a biopsy or use a CT scan to pinpoint the site of an infection. Chances are good that the death of Anne's uncle-brother escalated her facial pain, given the timing of the events and what scientists have observed about the consequences of intense negative emotions. But was her childhood trauma an original cause of her trigeminal neuralgia? There is a link between childhood trauma and chronic pain, but the complexity of interlocking causes and the difficulty teasing them apart are enormous. The head of Munich's Technical University's psychosomatic department had told me that the life stressors of his chronic pain patients tend not to be the original cause of their pain, but instead part of the reason an otherwise temporary event becomes an agonizing chronic problem.

Seemingly psychosomatic symptoms could also turn out to have nonbrain, physical causes. Several years ago, fibromyalgia patients cheered when studies showed that perhaps as many as 50 percent of them had damage to the tiny nerve fibers located under their skin, suggesting that their widespread pain wasn't a product of their brains after all. But it's still unclear whether this peripheral defect is a cause or an effect of the disorder, or whether it's a different disorder altogether. Chronic fatigue syndrome, a mysterious condition in which people feel pathologically tired for no apparent reason, used to be thought of as a psychosomatic condition (in the nineteenth century it was called nervous exhaustion), but researchers are now investigating how, for some patients at least, disordered immune cells and patterns of gut bacteria might be causing an ongoing flu-like immune response. "Theories that diseases are caused by mental states and can be cured by willpower are always an index of how much is not understood about the physical terrain of the disease," Susan Sontag said in *Illness as Metaphor*. Sontag wrote this as she was being treated for breast cancer in the 1970s, when it was common to hear the term *cancer personality* applied to a resigned, repressed, unfulfilled person. Tuberculosis was similarly thought to have an emotional link—overly sensitive people were more prone to get it— until 1882, when the German doctor Robert Koch discovered the bacteria that causes it. More recently, Arthur Shapiro, the New York psychiatrist who wrote about placebos, successfully advanced the idea that Tourette's syndrome is not a case of repressed thoughts and emotions violently pushing their way out, as some had believed, but a neurological disease.

Yet the inverse problem can also arise. A person's psychosomatic symptoms could launch an exhaustive diagnostic search that reveals an ovarian cyst, the misplaced tissue of endometriosis, the deterioration of spinal disks, a misalignment of jawbones, or any number of other red herrings that aren't actually producing any symptoms.

Chances are we will eventually arrive at a more refined understanding of psychosomatic symptoms, how they stem from a complex

conversation between the mind and the body. Research on chronic
pain conditions, panic disorder, and fatigue in breast cancer survivors all suggests that these problems often start somewhere in our
flesh—an injury, an infection, a rise in blood CO_2 levels, a cancer
treatment—but endure because of the brain's conscious and unconscious response to them. Does the incident pass and we move on, or
is our brain wired, whether for psychological or genetic reasons, to
overreact to it?

Before leaving Tübingen, I Skype with Paul Enck, a psychologist
who specialized in gastroenterology and surgery before getting into
psychosomatic medicine. He's been at the University of Tübingen
for two decades, but currently lives in Berlin. Enck, who has a big,
bald forehead, a white beard, and a jocular manner, tells me a story
about something that happened to him fifteen years ago that he still
vividly remembers. It started one day with severe pains in his lower
abdomen. It didn't feel like food poisoning or the flu because the sensations seemed to be coming from a specific area inside his intestines.
"It was the most frightening pain I had ever experienced," he says.
"For days, it made me depressive because it seemed like something
was really wrong." A gastroenterologist did an ultrasound and found
that Enck had a single bulging pouch in the wall of his colon, a diverticulum that had become inflamed and infected. He was given antibiotics, which cleared up the infection and got rid of the pain. But
Enck remained worried. His doctor said that if it happened again,
Enck might be a candidate for surgery, which meant doctors would
resect part of his colon. He was also terrified of the pain's returning.
"I've been in medicine for a long time, so I didn't think I would get
frightened easily, but for months afterwards I was running around
saying if this occurs again, they'll take out part of my colon because
that is what the guidelines say. I was even carrying medicines around
in case it happened again."

Then he paused to consider what was happening. "I said to myself,
'This is totally neurotic what you're doing. This is how you become
a patient.' I told myself, 'You're just anxious about getting the pain

again.' I had to get out of that loop of thinking there was something seriously sick in my body."

And that was that. Enck stopped the hand-wringing, his diverticulum grew no more inflamed, and his abdominal pain never came back. Maybe it was never going to. He will never know what might have happened if he hadn't changed his thinking. But as a veteran of psychosomatic medicine, Enck knows full well that the views we harbor about our bodies can be uncommonly persuasive, sometimes causing otherwise minor or temporary sensations to sprout into things that propel you to a doctor's office. Maybe vigilantly monitoring every little ache and pain in his gut and evaluating every negative sensation would over time have led to the unspecific gastrointestinal pain of irritable bowel syndrome or even, thanks to all the stress, another episode of inflammation. Enck is grateful he shifted his outlook so he never had to find out.

But what if one's mind-set is even more powerful than this? Everything we've learned up until this point argues against the idea. We've seen how our minds can shift the course of psychosomatic symptoms, change the illness experience of a disease, and lessen inflammation, all of which can add up to a world of healing. But in altering one of the worst physical injuries you can be subjected to, I didn't think the mind could play nearly as big of a role as it turns out it can.

Something to Believe In

Waking up neurons after a spinal cord injury

The gates of SaddleBrooke Ranch open as if into a promotional brochure for the good life. The morning I drive through, the air is cool and crisp, and the folds of the mountains are smeared pink and orange. All the houses here, in this luxury retirement community for "active adults" an hour north of Tucson, in the high Arizona desert, are large and tidy, with three-car garages and perfectly groomed yards. There seems to be little incentive to stay in them, though. Residents of SaddleBrooke Ranch have a dazzling assortment of socially stimulating and health-promoting leisure pursuits with which to occupy their days. There are sixteen pickleball courts, a creative arts center, a softball field, tennis courts, a massive fitness center with the casual elegance of a Four Seasons resort, and of course golf.

I won't be playing pickleball today, but I am going to a qigong class that starts at 9:00 a.m. in the Espíritu de Vida meditation room in the fitness center. This small room is just off the lobby, past the amber flames wobbling in the stone fireplace and the chatty line at the coffee bar. When I arrive, everyone in the group—a dozen women and one man—is assembled and ready to go. Our teacher arrives a few minutes later. He looks like the sort for whom pull-ups would barely get his heart going. Wearing black sweatpants gathered at the ankle and a black version of what look like nurse's sneakers, he is sinewy and compact. As he walks to the front of the room, his ankles and knees don't so much bend as uncoil themselves with feline precision.

Joe Pinella starts the class by explaining that the ancient Chinese practice of qigong, which I've never before done, is all about focused movements and strengthening our overlooked small muscles, along with the fasciae we barely realize exists. He says the Chinese refer to fascia—the thin sheath of fibrous connective tissue covering the entire body—as silk.

"We are building muscles the way a baby builds them," he says, his gray hair combed back and his blue eyes shimmering, "not tearing and building like we're at the gym. We're stressing and building gently and working the whole body to bring it into balance."

First, Joe has us do a series of deep breaths as we slowly raise our arms over our head and bring them back down. We also reach down to the ground and imagine bringing "beautiful energy" up from the earth. Then we do slow, deliberate turns of the neck and shoulder, also synced with breath, and I conclude this is going to be quite an easy class and I needn't have shown up raring to go in yoga pants and a tank top. But as soon as we start doing an exercise called teacups, my optimism fades. Joe has us bend forward and swoop our hands back and forth on either side of our body in a steady circular motion, all the while keep our palms facing upward. There is to be no wrist flipping, he says, as we bring our hands in front of us and then behind us again. Joe is demonstrating this effortlessly, like a bird barely flapping its wings against the horizon. Everyone else, though, seems to be having trouble. My teacups are flying all over the room's plushy iridescent carpet, and I feel like a kid trying to rub my stomach and pat my head at the same time. My brain can't make heads or tails of these strange routes and contorted angles. Nearby, one woman groans. "What am I doing wrong?"

Joe circulates around the room to help, and by time we're done, I bask in my success at engineering a few rotations with intact teacups. When Joe says, "Break," we all lunge for our water bottles. My back is covered in sweat and my shoulders ache from what apparently is not a cream-puff exercise for old people. I notice my arm joints, amid their fatigue, feel oddly emboldened, as if I have given them a set of angles they had never before considered.

Today Joe will teach two more classes and then practice qigong for at least an hour on his own. Now sixty-seven, he says he was once a quadriplegic, unable to move his arms or legs and confined to a wheelchair after a horrific car accident. That he is not still in a wheelchair or using a cane or bearing any obvious signs of his previous injury has everything to do, he says, with qigong, which he used to slowly bring himself back to full health, over seven long years. After the class, over lunch at a nearby Mexican restaurant, Joe tells me about this improbable scenario.

The accident was on a brilliantly blue June day in 1991. He and his then wife were heading back to their hometown in New Jersey for an extended visit. Joe had packed up his Bronco with several large suitcases and the couple's three German shepherds. He had also attached a new, custom-designed trailer that housed his shiny red Ferrari. At home in Bayonne, before moving out to Arizona a year earlier, cars had been Joe's life. He had owned auto body shops that dealt exclusively in luxury cars such as Ferraris, Lamborghinis, Rolls-Royces, and Aston Martins, and he'd started several businesses that sold and insured them. The work he did on these cars was painstaking. "I used to spend a hundred dollars for a gallon of German primer instead of fifteen dollars in the US," Joe says. "People would say I was wasting my time being so exacting. I'd say, 'Okay, I'll waste more time.' I guess I'm pretty OCD and type A. At one point I had twelve businesses and worked ninety hours a week."

With the Ferrari inside the trailer, Joe eased the rig out of his manicured Scottsdale neighborhood and onto the northbound lanes of Interstate 17. When he reached the juncture of Interstate 40 just outside Flagstaff, Joe suppressed the urge to gun it around the two semitrucks he'd been riding between. Not many days were better for driving than this one, and he decided to savor the moment. Perfectly clear skies, low humidity, zero threat of rain.

When, a few minutes later, the trailer hit the Bronco in a thunderous crunch, Joe didn't realize what had happened. It was so fast and unexpected that records would later show he didn't even slam on the brakes. The trailer had come undone on a downhill stretch and shoved the Bronco off the road in a series of somersaults down what was apparently a two-hundred-foot-deep ravine. Before he lost consciousness, Joe remembers a man leaning over him asking, "Can you move your arms or legs?" Joe considered this for a moment and then replied weakly that he couldn't. "Call Flight For Life. Tell them we've got a quad here," yelled the off-duty paramedic who had been riding in the semi behind Joe.

Joe breaks off in his story. "I really don't like to remember. Sometimes these things can draw families closer together, but sometimes it pulls them apart." We pause to focus on our food, which has quickly arrived at our table in the mostly empty restaurant. A vegetarian since high school, Joe has ordered the veggie burrito and uses his fork to scoop out the veggies, leaving a deflated tortilla abandoned on his plate.

After a few minutes, he continues, describing how the force of his head against the top of the car ruptured a series of disks in his neck and forced his sixth cervical vertebra, or C6, to jump out of position and land almost on top of the C5 vertebra above it. He was quickly taken to the trauma unit at the nearby Flagstaff Memorial ER, where a neck-and-spine specialist and a neurosurgeon were called in. Because, generally speaking, the higher up on a spine an injury is, the more severe the consequences and the greater the loss of movement, Joe's neck trauma was among the more dire of scenarios. A portion of his upper spinal cord, which runs through the protected channel in the middle of the spine, had been injured and was blocking all nerve signals going into and coming out of his brain. When he first regained consciousness in the hospital, Joe says he couldn't move or feel anything below his neck. His wife's injuries were to her lower back, hips, and legs and, though also severe, did not affect her spinal cord.

In the hospital—Joe thinks it was about a week after the accident—a surgeon ran what's called a pinwheel over his legs and Joe reacted in pain. "It looked like a tiny pizza cutter," Joe remembers. "And it felt like he had just pricked me." At first the surgeon, Dr. Donald Hales, didn't believe that Joe could really feel the device. Dr. Hales assumed Joe was seeing the wheel move over his leg and remembering what it would feel like. So he made Joe close his eyes. But Joe still felt the wheel, which meant that some sensory signals were making it through the injured area of his spinal cord and into his brain, giving greater hope for at least some degree of recovery. Dr. Hales immediately ordered several more operations. His team removed pieces of chipped vertebrae from the surrounding tissue of the neck, and he and the neurosurgeon performed multiple procedures to stabilize Joe's neck. They harvested a small section of each hip bone, carefully molded the pieces into the shape of two vertebrae, and then replaced the two shattered ones with them. At the back of Joe's neck, they fused those vertebrae together with two metal plates and a set of four screws.

During the four weeks Joe was at Flagstaff Memorial, now called the Flagstaff Medical Center, people who lived near the scene of the accident showed up periodically with retrieved clothes and luggage. Eventually, most of Joe and his wife's possessions found their way back to them, as did the two adult dogs riding in the backseat. But for Joe, getting news that his puppy hadn't made it was almost too much to bear. He wasn't grieving the dog so much as that its crate had, at the last minute, taken the spot where his twelve-year-old stepdaughter was supposed to have sat. In an auspicious change of plans, she had taken a flight to Newark the night before.

Amid his distress, Joe found an odd and unexpected thread of hope. While unconscious and undergoing surgery, he says he experienced the sensation of hovering over his body and looking down upon the scene below. When he awoke, he had an unexplainable and wholly unreasonable sense of optimism. Dr. Hales had given Joe a "thirty million to one" chance of ever returning to a normal life, but Joe did not believe this. He'd recovered from injuries before—from daredevil skiing, high school gymnastics, and even a previous car accident—so why did this need to be that different? Such utopian musings were not well received by the hospital staff. To them, Joe seemed to be sliding into a deep depression, beyond what was normal for a person in this situation. "They all said, 'You're refusing to face reality,'" he recalls. He was prescribed antidepressants.

A similar reaction unfolded at the physical therapy office Joe was referred to after he left the hospital. The therapists there saw it as their job to help severely injured patients prepare for the possibility of living the rest of their lives in a wheelchair. They surmised that even if Joe regained some movement in his arms and legs, he was highly unlikely to ever lead a normal, independent life, and they wanted to help him adjust to this new, handicapped reality. They talked of motorized wheelchairs and avoiding bedsores.

Still feeling irrationally optimistic, Joe quit physical therapy after just a few months and began an unorthodox and lonely period of his life, which started one day with a phone call. With a friend pressing the phone to his ear, Joe reached a Buddhist temple in New York City that he had frequented during high school. One of Joe's friends had an older stepbrother who'd moved to Taiwan to become a Buddhist monk, and while home visiting one summer, the stepbrother had brought the two boys along to Chinatown. Joe remembers watching intently as monks meditated, chanted, and breathed heavily while doing slow, deliberate exercises with a grace and strength he'd never before seen anyone move with.

"They would stand on one leg and do a squat, putting their butt on the ground, and then stand back up. Every one of those monks could do that. It was just part of their practice. They didn't think of it as being strong." When the stepbrother returned to Taiwan, he asked if the sixteen-year-old Italian kid from Bayonne could stay on and learn what Joe later came to identify as qigong. He trained with the non-English-speaking monks for several years until he got busy with work in his twenties. After that, Joe didn't think too much

about the temple. But now that his body was failing him, he remembered it had been a kind of unlicensed health clinic for the people of
Chinatown, a place where monks treated people with herbs, acupuncture, qigong, and a kind of energy healing called medical or external
qigong, in which people are said to emit qi from their hands to heal.
(Joe himself never saw anyone there do such a thing.)

When Joe called the temple, it took a few minutes to reach someone who spoke English. "Tell them it's Joe Pinella and I studied there
in the sixties. I've been in an accident and broken my neck. I want to
see if qigong can help."

A few days later the one man at the place who spoke English called
back. "The master says it's no problem. Qigong is just what you need
to help heal your body."

"You don't understand," Joe replied. "I can't move. I'm paralyzed."

The man said he would look into this. A few more days passed
and then there was a call, this time with more pointed instructions.
The mind directs the qi, the monk said. Use your mind; trick it into
believing that you're a baby again, growing in your mother's womb.
Do this every day. Get into a warm swimming pool, heated to 98.6
degrees (the master was very specific on this), and shut your eyes.
Imagine building a heart, a brain, lungs, and nerves. Visualize your
nervous system using an alternate route around your injuries. Imagine *qi* moving through your body.

The monk also recommended that Joe learn about his injuries
from an anatomy book. So Joe got himself a copy of *Gray's Anatomy*,
the nineteenth-century textbook that would later spawn a TV show,
and studied how the brain, spinal column, nerves, and muscles are
all linked together. He had no idea whether such elaborate visualizations would work, but he knew it was worth a try, especially since he
had a hot tub in his backyard. So twice a day, home health-care aides
wheeled Joe out into the backyard, set a plastic chair in the middle of
the tub, and hoisted him in.

"I sat for two or three hours at a time with this big straw hat
on," he says. "I would close my eyes and then just imagine, imagine,

imagine. I never wanted to give myself the feedback that it wasn't working. If you imagine that you're moving something, then that's what the body believes."

He did an exercise called compression and expansion, in which he would take a large breath and imagine squeezing in his belly button toward his spine, arching his toes up toward his head, and spreading his fingers out as far as they could go. Then he'd release it all and breathe out. Even though nothing was actually moving, Joe started to feel a dynamic pulsing sensation when he'd release. "I knew it was just the movement of the breath, but I started to feel like I could move my whole body." Instead of dismissing it as inconsequential, Joe took it as an encouraging sign. "It gave me something to work with." He also imagined bright white light—qi, he says—moving from his head and down through his whole body, exiting through his toes and fingertips.

Joe decided, after at least four months, to open his eyes to see if anything was happening. When he did this, he was stunned to notice fingers wiggling on his right hand, the first time anything below his neck had moved on its own since the accident. This meant a motor signal had traveled down from his brain and through the damaged area of the spinal cord, and if this had happened, Joe reasoned, maybe other signals could come through, too.

Not long after this, with his arms out in front of him in the bubbling froth of the hot tub, he noticed that if he looked to one side, his arms would slowly migrate in that direction, almost as if he could direct them to move with his eyes. "It was interesting that I could get both of them to go to one side and then the other side. So I just started to wave back and forth."

From there, Joe says he followed his pain. When he'd start to feel tingling and sharp pangs, he took it as an indication that, as the monk had predicted, qi was starting to move. He'd focus on where the sensations were coming from and try to get the feeling to spread, to get more qi moving. Using his newly acquired understanding of anatomy, he also imagined nerve signals using other pathways around the damaged area of his spinal cord. By the end of the first year, he had a good deal of mobility in his legs and was shuffling around for brief spurts with a walker custom-designed with a hook for his left arm, which still had no sensation and was prone to violent spasms.

Such standing and shuffling was easy, Joe says, compared to relearning all the basic fine motor skills his body no longer knew how to do. "Nothing was controlled at that point. My arms and legs, everything was flopping around like a fish. When I'd try to feed myself, I'd go like this." He picks up his fork and sails it past his mouth. Food would end up on the floor, on his lap, and strewn across the table. "I'd watch how my stepdaughter ate and then try to break it down into separate motions and then slowly combine them. That's what we do in qigong. You break down postures into separate moves. Everything I give to people in my classes comes from personal experience where I had to slowly and methodically learn to walk again, to balance again, to get up again off the toilet or a chair. For a lot of older people that's a big deal. And it certainly was for me."

Joe's progress, encouraging as it was, did little to quell the growing storm between him and his wife, who was dealing with her injuries in far more conventional ways. She never understood Joe's decision to stop working with physical therapists and doctors. It seemed reckless, not unlike his decision to tow a hot rod on a trailer that had obviously not been secure. (It was later determined that the welding on the hitch had air pockets.)

"We had two really different ways of looking at how our lives were going to be after the accident," Joe says, his jaw tensing. Three years after the accident, he and his wife split up acrimoniously. Joe isn't in touch with her and hasn't spoken to his stepdaughter in twenty years.

The first few years after his wife left him were extremely trying. "I would get depressed and quit for a while. But it wasn't like I could go out and drive somewhere or go for a walk or go bowling or to the mall or anything. I'd just sit there and feel like killing myself. But I'd go back to qigong because I always felt better when I was doing it." He would also summon the uncanny sense of trust he had felt in the

ER and focus on something one of the monks had taught him at the temple. One day, a monk had told him to stand on a piece of paper lying on the floor and asked, "How much taller do you feel?"

Joe replied that he didn't feel any taller. The monk put a book on the floor. "Stand on this. Now how much taller?"

"Umm, maybe an inch or two."

The monk opened the book and showed Joe that it had 365 pages. "Every time you do this practice, you're getting one slip of paper, but you're not going to notice how much better you're getting."

Joe, his current wife, Debi, and I are all standing in the parking lot now. Joe is heading off to teach another class at another retirement community, and I go back to my hotel. Barreling down a lonely stretch of desert highway, I marvel at how complete Joe's recovery seems to be. He doesn't appear to have any awkward movements or tremors or anything else you'd expect from someone with former paralysis. The next day, when we meet up again, Joe tells me that a few things do bother him. When his neck gets cold, pain shoots down his back, and he still can't write well with a pen or pencil. More recently, "sparkles" of pain have appeared in his neck when he turns it from side to side, and occasionally out of nowhere a "hot needle" sensation emerges from his shoulder. Sometimes, his left arm and middle finger will also go numb, and Debi says that at night his arms and legs occasionally shake and twitch uncontrollably.

These developments have Joe worried enough that he recently got some X-rays from a chiropractor and scheduled an MRI with a neurologist. Over another lunch, he hands me copies of the X-rays, which were done a few days ago. They show a black mass of metal hovering over a hazy outline of bones. Peering closer, I see a square plate and four screws, which look like pieces of an Erector set. "You see this screw here that's out of place?" Joe points to one that's slightly ajar. "No one knows what the tip of that screw is doing. Is it coming loose and pressing on a nerve?" (Later, a neurologist will tell him that the screw is probably encased in bone and not moving or otherwise causing problems.)

"I start to think, 'Maybe this is it, I've reached a peak, and I'm going back. What a great twenty years I've had, and now with age, I start to decline.' It scares me. I've started sleeping in a neck brace again. I'm thinking that something could happen when I sleep, that I might pinch a nerve or something."

Toward the end of our conversation, Joe gets emotional again, at one point thinking back to that June day and wondering aloud, "Joe, why didn't you just turn the wheel, why didn't you just turn the wheel?" I find myself trying to console him by emphasizing the hope and optimism that obviously drove his recovery. I remind him that the belief in his body's capacity for self-healing is what set him off on this path and fueled all those hours trying to figure out how to get off a toilet and curl fingers around a spoon. But such assurances instantly feel hollow because the truth is that I have no idea how all the different things Joe did—his elaborate visualizations, his positive response to pain and other bodily sensations, his irrational exuberance, and his repetition of small, deliberate bodily movements—could possibly have taken him from quadriplegia, or tetraplegia as it's now called, to Jack LaLanne. We've heard how exercise done with a positive mind-set can change how the brain processes bodily signals and thus help ease pain. But Joe didn't have only musculoskeletal pain, he had a damaged spinal cord, too. How could any of these things have healed someone of paralysis?

I am old enough that I remember Christopher Reeve falling off a horse, injuring his spine (two vertebrae up from Joe's injury), and then, after years of well-funded therapy, dying an incapacitated fixture on a wheelchair, the great Superman only able to wiggle his fingers. This may contribute to the impression I have that a spinal cord injury can't heal itself the way a broken bone or torn muscle does. But this view isn't entirely true, as I learn from a 2007 paper written by a panel of twenty scientists who evaluated spontaneous recovery after spinal cord injuries. Looking at clinical trial data from seven studies totaling roughly two thousand spinal cord injury patients, they found that many people do not remain trapped in the same miserable state they were in at the ER.

"Most spontaneous recovery from spinal cord injuries occurs within the first three months," the authors wrote in the journal *Spinal Cord*. "After that, small amounts can take place for up to 18 months and occasionally beyond." Roughly 20 percent of people with the most severe, or ASIA A, injuries, they said, experience some recovery of sensation or movement, and up to 40 percent of those with ASIA B gain additional sensation and movement, mostly within the first year. In a classification system developed by the American Spinal Injury Association, an ASIA A rating means no movement or sensation whatsoever below the injury and the lowest expectations for recovery. Those with ASIA B have some sensation but no movement below the level of the injury. ASIA C patients have some motor control in at least half their muscles. D means even greater control in these muscles, and E is moving and feeling in all the major areas of your body.

We don't know how Joe's injury was classified while he was in the hospital because no medical records are available. Any documents Joe once possessed were allegedly thrown out during his messy divorce and the exodus from the house he and his first wife had shared. Debi says she has made multiple attempts in recent years to locate records, only to be told that they no longer exist due to the routine purging of medical files. When I contacted the Flagstaff Medical Center, they told me more or less the same thing. I was given a record number with Joe's name and the year 1991 on it, but it was empty. After ten years, I'm told, files are destroyed unless someone is a current patient, and Joe never went back after being discharged. I encounter a similar story at the Arizona Department of Public Safety, whose officers patrol the highways. They have a record of Joe's accident occurring on June 5, 1991, but all the details are gone.

But based on Joe's recollections, which I corroborate with several friends who visited him in the hospital, and on conversations with

Dr. Donald Hales, the spine surgeon, whose memory hasn't been so methodically purged, we can conclude that Joe is likely to have been discharged with an ASIA B spinal cord injury. Dr. Hales, who was Joe's primary doctor for those four weeks in the hospital, tells me he remembers Joe not being able to move anything below his neck but having some degree of sensation. That Dr. Hales recalls Joe at all after twenty-five years is a function, he says, of the fact that a Ferrari tumbled off the highway and the severity of his injuries.

"I recall being called to the emergency room to see him," Dr. Hales says. "It was pretty apparent that he had a broken neck, and that was confirmed on X-rays and a CT scan that we obtained. We call it jumped bilateral facet joints. He had both his motor and initially his sensory out, and I would not have expected him to have a full return of both, much less be teaching and training in a gym.

"That's an amazing level of function for someone who was a quadriplegic. Even when people make what you would consider a remarkable recovery, they're not normal. So it really is one of the unusual cases. In a church setting you might call it a miracle or faith healing. Outside of church I think we tend to call it positive mental attitude."

It's almost certainly more than that. It wouldn't be a stretch to conclude that Joe was among those 40 percent of ASIA B patients lucky enough to regain spontaneously some degree of sensation and movement. Perhaps it produced some of his early finger and toe wiggles and those hovering arm movements, as well as the pain sensations he felt as some of his nerves came back to life. But such natural healing didn't give Joe the ability to walk normally, feed and bathe himself, drive a car, and teach a dozen weekly qigong and tai chi classes.

For that we have to look at what Joe did in the name of rehabilitation, which recent spinal cord research has revealed to be a more attainable goal than anyone ever thought. The first sign that something novel was underfoot came a decade ago at the Kentucky Spinal Cord Research Center in Louisville. A team of scientists from there, UCLA, the California Institute of Technology, and the Pavlov Institute of Physiology in Saint Petersburg surgically implanted a tiny electrical stimulator onto the spinal cord of a twenty-five-year-old man, Rob Summers, who had been hit by a car three and a half years earlier. Much like Joe when he left the hospital, Summers could feel some sensation in his legs but was paralyzed from the waist down and confined to a wheelchair. Summers had diligently spent two years working with the Louisville therapist staff trying to walk. Yet without others helping him, he still couldn't move his legs or stand up.

But now with a stimulator delivering small bursts of low-level electrical current directly into his spinal cord just below the site of his injury, Summers did something unprecedented. Positioned amid an apparatus of horizontal bars and a bungee-corded harness, he awkwardly hoisted himself out of his wheelchair and was steadied by therapists for several minutes. When the therapists let go, Summers stood there, to everyone's amazement, bearing 65 percent of his own weight for the first time since his accident. He remained like this for four minutes while scientists watched EMG monitors record contractions ricocheting through his leg muscles. About seven months later, with the stimulator turned on, Summers was able to intentionally create movement in his toes, ankles, knees, and hips. So too were three other men who were later implanted with stimulators, including one with an ASIA A injury who had been told there was no hope of any recovery. After four to eleven days of stimulation, each of these patients could initiate movement in his legs.

"This was a huge surprise," Dr. Reggie Edgerton, the principal researcher from UCLA, tells me. "Because no one thought you could regain function across a complete lesion." By "complete lesion," Edgerton is referring to ASIA A injuries. "What everyone always assumed was that in a spinal cord lesion, all the cells die. But I think it's now obvious that some of the cells get damaged but don't die. There's connectivity even though there's no function. That's a concept that's never been considered before."

Edgerton, who has been doing spinal cord research at UCLA

for three decades, believes that the continuous jolts of electricity from the spinal cord stimulators "woke up" dormant neurons and prompted them to grow new connections to other neurons, thereby creating communication pathways. "I think this idea of spinal cord plasticity has alerted everyone to the idea of greater possibilities. The standard program for a spinal cord injury has been that you get a couple months of rehab if you're lucky, and then you go home and that's it. This new thinking gives hope to people that really had no hope of recovery."

But Edgerton cautions this is still hardly a Holy Grail for the estimated 250,000 to 1.3 million Americans living with a spinal cord injury. None of the men he's worked with are lurching out of their wheelchairs to walk on their own. (Eighty percent of all spinal cord injuries happen to men, who find themselves in physically perilous situations far more often than women.) They've gotten control back over certain autonomic functions, such as their body temperature, blood pressure, bowels, and bladder. But without the electrodes pumping out currents, most of their knee bends and leg kicks don't happen. What's more, they can do only one side at a time. You have to turn the epidural stimulator off and then back on to activate another set of muscles.

Other research tackles a different piece of the movement puzzle. At the University of Pittsburgh School of Medicine, scientists decided to bypass the spinal cord entirely and head straight for the brain, the body's command center for movement. In 2012, neurobiologist Andrew Schwartz and his team did brain surgery to place two microelectrodes, each a quarter of an inch in size, into the motor cortex of a fifty-two-year-old woman who had been paralyzed below the neck for thirteen years. Jan Scheuermann's paralysis came from spinocerebellar degeneration, a condition in which the brain's ability to relay signals to the muscles gradually and inexplicably goes away. Located just above Scheuermann's left ear, the electrodes record the signals that between sixty and two hundred of her individual neurons generate while she is thinking about moving her arms. Schwartz then takes these intentions, or "action potentials," decodes them with software, and feeds them into a robotic arm, a feat that should put to rest any notion that our mind's thoughts and intentions are not discernible biological things. With her thoughts animating the robotic arm, Scheuermann is able to pick up objects and feed herself, although a few months of focusing her thoughts and practicing movement were needed before she was able to master such fine motor movements.

In Switzerland, Greg Courtine is trying to combine both approaches—Schwartz's so-called brain-machine interface and Edgerton's spinal cord stimulations—to regain movement with actual limbs, not robotic ones. He's already done it successfully in monkeys with a paralyzed leg. Electrodes implanted into the monkey's brain wirelessly beam its intentions about moving its leg to a stimulator placed in its spinal cord below the level of injury.

"The monkey is on the treadmill walking with its three legs until he thinks about using that paralyzed leg, which is something we can see from monitoring brain activity. When he thinks about it, the spinal stimulation turns on and he can use that leg. It's completely crazy," Courtine marvels. "We still don't fully understand the mechanism, how the stimulation is integrated and cooperates within the natural activity of the brain." A professor at the Center for Neuroprosthetics and the Brain Mind Institute in Lausanne, Courtine says he is in the beginning stages of testing the approach in humans.

I think back to Joe sitting for hours, raisin-skinned in his hot tub, pretending to move his body in precise ways and trying to send nerve signals into pathways around injured areas and then, once some movement returned, doing slow, repetitive, mindful qigong-based motions in his limbs. It's been known for some time that visualizing actions, whether kicking a soccer ball or playing the guitar, generates almost the same brain activity as actually doing it, so I wonder what effect Joe's flurry of electrochemical brain activity could have had. "Can imagining movement produce signals that go from the brain down into the spinal cord?" I ask Courtine.

"Sure," he replies. "We do know that spinal circuits get modified during imagined movements."

When I tell him some of the details of Joe's case, he doesn't react with skepticism. "If it's a complete injury with no motion or sensory, then no, I don't believe that dramatic recovery is possible for someone on their own. But otherwise, I think it's completely possible. I've seen it in my lab in rats. It's remarkable how much effect the willpower of training can have on their neural systems, so imagine humans with a lot of drive. Although if someone would get stimulators implanted into their spinal cord, it makes regaining movement a lot easier."

Reggie Edgerton also thinks it's feasible that years of repetitive thoughts and movements could be their own kind of electrical stimulation and reanimate sleeping spinal cord neurons. "We haven't studied those kinds of individuals, but I don't question it at all. My view is that what's happening is that if there's some remaining function and if the person is determined enough to remain active, they can push that circuitry to change in ways that we know it can. It's remarkable how little connectivity you need in order to regain a lot of function. But you have to give it time. If you don't engage these circuitries, they don't improve."

"Engaging circuitries" is exactly what many leading spinal cord rehabilitation centers now seek to do with patients, regardless of the level or severity of their paralysis. At the Kennedy Krieger Institute's International Center for Spinal Cord Injury in Baltimore, director Dr. Cristina Sadowsky says she uses a number of devices to try to regenerate lost sensation and movement. There are treadmills with supporting harnesses and robotic exoskeletons for walking, stationary bikes with electrodes that attach to leg muscles to electrically stimulate them, and, interestingly, submerged treadmills in pools that are heated to between ninety-two and ninety-four degrees to minimize muscle spasms—a few notches cooler than what the monk recommended for Joe. "If someone has no movement on land, maybe they're in the pool with a floatie and imagining moving. The idea is to take away gravity so it's going to make it easier for someone to get that incremental motion," Sadowsky says.

But regardless of which device is used, repetition is key. "We have people do 'targeted activity' an average of about four hours a day, five days a week, for at least two to four weeks. After that you don't need to do it at the same level, but you still have to do it."

Kennedy Krieger patients also do a form of the mental focus and body awareness Joe practiced. "We always ask the patient to think about what they're doing when they're on the bike or walking," Sadowsky says. "If I'm not visualizing and sensing the muscles contracting and the limb going in one direction, then I'm not really focusing on restoring transmissions through the damaged area. You have to connect the limbs through the spinal cord to the brain's motor cortex. If you're walking, figure out where your foot and hip is in space, paying attention to the form."

She says that those with an ASIA B injury who do all this repeatedly and consistently have at least a 50 percent chance of walking unassisted one year after their injury, with higher chances the earlier you start. But contrary to popular assumption, such "activity-based restorative therapy" can still be beneficial even a year or more after an injury. "Most people think that at one year if you're not walking, you have a poor prognosis, but that's not exactly how we look at it. It's not black and white."

It's startling to think of Joe's approach, originally supplied by a qigong master in Chinatown twenty-seven years ago, as a low-tech version of many of the things now known to be effective for spinal cord injury rehabilitation. Joe didn't have electrodes implanted into his spinal cord or brain signals beamed to a robotic limb, but for seven long years, often in a warm and low-gravity environment, he practiced repeatedly stimulating his brain's motor circuitry, his spinal cord, and the nerves leading into and out of it. He did this day in, day out with a mix of imagination and small, slow, mindful muscle movements and exercises that he combined into more coordinated maneuvers. All because he stubbornly believed that the ancient practice of qigong could awaken his nervous system, even though the conventional thinking at the time had told him quite the opposite— that there could be little nervous system plasticity following severe spinal cord injuries.

This doesn't suggest qigong as an effective therapy for spinal cord injuries. Maybe it is, maybe it isn't. It certainly seems to have worked for Joe, and it's possible it might be helpful for others. But without large, high-quality clinical trials, we simply can't draw any definitive conclusions. Outside of a few small studies in China, qigong and tai chi have never been studied as treatments for spinal cord paralysis, and as far as I know, no spinal cord center is using them. But tai chi has been proven effective for certain types of motor control in Parkinson's patients and the elderly, suggesting that it might be helpful for spinal cord injury patients struggling to regain coordinated, fluid movement.

Joe's story does tell us definitively that, while mind-body healing has limits, in some cases a deep desire to heal oneself may push those borders further out. Joe's views steered him toward constructive behaviors that over time made profound differences to his health. His mind probably didn't repair the lesion in his spinal cord. But without his buoyancy, focus, obsessiveness, and, arguably, his naivete, he would never have found his way around that lesion to those spinal neurons that could still be awakened.

Joe was also right that the doctors didn't know everything. More is known on the subject now, but even more is likely still to be learned about the mind-body connection in this arena. But the arena matters: Joe had the advantage, if you can call it that, of trying to undo a finite injury, instead of working against an active disease process. Trying to engage the nervous system to eradicate degenerative neurological diseases such as Parkinson's, multiple sclerosis, Alzheimer's, or a brain tumor is going to present very different challenges.

Joe says he now works occasionally with people who have such neurological problems, to help them gain a combination of strength, stability, and mobility. "When people have something they can't do, I say, 'Okay, try it this way.' There's never a no," he tells me. "There's a woman with MS that I've been working with for a little over a year. She calls me two weeks ago from the office of her physical therapist, who she's been working with for fourteen years. She wants to tell me she did five sets of leg thrusts from a squatting position. When I met her, she could hardly walk."

Once in a while, people who have paralysis from a spinal cord injury also call Joe. He doesn't make promises, but still won't say no. "They ask me if I can help them, and I tell them I never know. All I can do is tell you my story and show you what I did. Maybe something can happen."

14

· · · ·

Believing Is Seeing

Searching for signs of healing energy

For Joe it was qi flowing through every system of his body. For Danila Castelli, the healing hand of God had touched her. Gloria McCahill said she had helped repattern Ian's blocked energies. Between Adam Engle and Peter Churchill, it was a sacred transmission of "higher energetic healing frequencies." Everyone, it seemed, believed in healings that were the work, at least in part, of some ethereal or spiritual healing force. Most often the term was *energy*.

To me, it seemed in every case as if science could provide more probable explanations, with no need for one's feet to leave the ground. But at some points along the way I yearned to have a reason to believe. "Stay open-minded. 'I couldn't find a needle in the haystack' doesn't mean it isn't there," Donna Eden's husband, David Feinstein, had advised me. The possibility of phenomena that science has failed to detect has a tantalizing thrill. Throughout history, many ideas have been unreasonable and preposterous until they weren't. Gravity, electricity, the ability of incubators to keep premature babies alive, washing one's hands after examining cadavers—all notions laughed off as pseudoscience until more experiments proved them to be right. My sons have a science book that relates the derision German physicists ran into in the early 1900s as they tried to prove that atoms weren't the smallest particles in the universe and, furthermore, weren't really particles at all. It's not crazy to imagine that other such unreasonable ideas are out there still to be discovered. The strangeness of dark matter, the inner workings of human consciousness, and the activities of subatomic quantum particles, which can be in two places at the same time, certainly all suggest it. "The universe is full of magical things patiently waiting for our wits to grow sharper," the English author and poet Eden Phillpotts wrote.

So, the question: Is there such a thing as healing energy? I first surveyed attempts researchers have made over the years to capture, measure, or otherwise track it. A former UCLA physical therapy professor named Valerie Hunt, for instance, developed a BioEnergy Fields Monitor, which she claimed to be "the most valuable diagnostic tool ever developed." By monitoring the electrical properties of muscle groups, it supposedly mapped out not only the patterns of energy that preceded disease, but also had the capacity to prescribe the right healing approach for a particular person. Hunt's device was never submitted to other researchers for testing, however, because Hunt, who is now deceased, was the only one who could operate it. She was gathering up all the data coming from the probes and interpreting it herself, which isn't a particularly scientific way to go about things.

In North Carolina, at the Rhine Research Center, once a part of Duke University, John Kruth is a former software developer who works with the hypothesis that healing energy either *is* biophotons, the tiny, subatomic particles of light emanating from living cells, or can be tracked using them. He's put people in pitch-black rooms and measured their biophotons with powerful machines that amplify light. When certain healers, meditators, or martial artists have sat in the room and meditated or, as they put it, "run energy," their biophoton readings jumped four to fifty thousand times above what they were before and what the average person emits just sitting in the room. But this happens with only 10 percent of the healers Kruth brings in, a variation he says he doesn't yet understand. He also hasn't looked to see whether other types of consciousness-shifting activities can cause such biophoton spikes—such as a mathematician thinking about a complicated problem, an actor summoning an emotional experience, or a chess champion strategizing about a game.

Then there's William A. Tiller, a former head of Stanford's materials science and engineering program, who created an apparatus that can allegedly capture the focused intentions of practiced meditators. In a series of experiments, he put meditators into an electromagnetically shielded room and had them concentrate on a specific intention, such as changing the pH of a glass of purified water. During the meditation, the device was supposedly imbued with the specific intention. Tiller then placed the device near a container of purified water for a time, during which the "subtle energy exchange" between the device and the water resulted in the intended change to the pH of the water. Tiller, who was featured in the documentary *What the Bleep Do We Know!?*, claims this is a demonstration of intentions changing physical matter. But when I spoke with him, he wouldn't tell me much about how his device works. It also hasn't been subjected to any published, peer-reviewed testing.

Other researchers have the ostensibly more practical view that we should be looking for healing energy within the known electrical signals of the body. I get briefly excited after reading a 2015 paper that outlines such a search and features the name of a respected cell biologist/computer scientist among the authors. Michael Levin of Tufts University studies the bioelectrical signals that are produced by all cells, not just communicating neurons. These weak fields are already known to be involved in wound healing and bone repair, so maybe, I think, there's some evidence that these signals form a kind of coherent, whole-body healing force. Not so, says Levin. "There are little to no data indicating that somatic bioelectricity is closely related to flows of any putative mystical energy in the alternative therapies field," he writes in an email. "Certainly, there's no data from my work. Anything like that would be pure conjecture at this point in time."

In the absence of hard evidence, it seems what I may need to have is a personal encounter with the anomalous force so many others say they have experienced. Gloria McCahill told me that once when she was clearing Ian's aura, her middle and index fingers began to shake and vibrate as if being sucked into a vortex. On other occasions, she felt a sharp stabbing sensation in her hands. While working on someone in a darkened living room years ago, Peter Churchill had seen a "three-foot-wide shimmering sphere of golden white light" hovering in the air between the ceiling and the floor. It sent out twin streams of light toward him, and when these beams of energy entered him, he said it felt like "one hundred thousand volts of electricity." He was stone-cold sober. Many others told me about less dramatic but nonetheless unexpected feelings of warmth and tingling during Reiki, Eden Energy Medicine, or acupuncture treatments.

If I experienced something like this, it wouldn't prove anything. But it would force me to consider the limitations of my materialist worldview and give me a reason to keep searching. The laws of human narcissism dictate that while it's easy to explain away other people's unorthodox experiences, our own deserve to be taken more seriously. As a toe in the water, I sign up for an online "Seeing and Perceiving Life Force Energy" seminar run by a Scottish woman named Deborah Gair, who practices an energy healing method called Quantum-Touch. She promises that enrollees will "discover gifts you did not realize you had" and that the class will "bring your healing abilities and your own personal unfoldment into a whole new paradigm." She says that sensing energy is a matter of becoming more open and sensitive to it, and that such things can be taught, much like one might learn to meditate or walk a tightrope.

So for six Sundays in a row, I take my "unfoldment" and watch Deborah talk on a video feed that periodically freezes, causing the class to divert into a discussion about who can and can't see and perceive Deborah. When the technology is working, she explains that sensing life-force energy is not a matter of using our physical eyes and ears, at least not initially. We have to close our eyes and tune inward, use our inner eyes and ears. "We will be reading the only book that's worth reading, and that's you," she says with a rolling

Scottish accent and giant grin. "Fling away all the ideas you have about your body and get curious."

Deborah tells us that her own seeing and perceiving journey began thirteen years earlier and has produced "extraordinary things on a regular basis." This includes, she says, watching sparks of light emanate from her hands, seeing her legs dissolve into beams of light while in the shower one day, and seeing physical objects such as a small Buddha statue mysteriously materializing and then disappearing. "You have boundless potentials and possibilities that are quite literally beyond your wildest dreams," she says.

To help us get to these glorious new realms, we are to do fifteen to thirty minutes of guided meditation during each class, and ten to fifteen minutes every day during the week as homework. Speaking slowly and tenderly, Deborah tells us to "take deep breaths of fresh cosmic breath into your belly" and to "penetrate the darkness" behind our closed eyelids. In the first class, with my eyes shut and thoughts of Deborah's disappearing legs stuck in my head, I get the strange sensation that my body and mind are being disconnected. I feel as if my body could go off to sleep, as it yearns to right now, yet my mind would still be very much awake. The feeling doesn't last long, and when I mention it to Deborah on the live chat board we use to communicate during class, she takes it as a great sign. "You, as 'Melanie,' are far more than a body. Your body is actually in what you are. You are that and also waaaaaaay beyond."

Others, though, seem to be having far more interesting experiences than I am. During the meditation, Marcela from Florida sees a bunch of pulsating colors. Another person had a "tingling pressure" in her "third eye." The following week, someone else emails Deborah to say that as she was going to bed one night, she saw light spots hovering on the other side of her room. One of them moved toward her and transformed her hand into a white ray of light. "There's definitely a sense of clairvoyance developing," Deborah declares. Such reports, delivered so early in the class, seem odd. If people such as Marcela and Sally are already so adept at seeing and perceiving life-force energy, why do they need this class, which I had figured was for the more paranormally challenged such as myself?

On week five, we try our hand at seeing and perceiving with our eyes open. To do this, Deborah shows us several pictures on a screen, and with no additional information, we are to tap into our powers of clairvoyance to give her details about them. The first is an underwater photo of a scuba diver giving a thumbs-up. I wager a guess that this person has no kids (what parent has time to scuba dive?) and that he might work in sales (he seems cheery), both of which, Deborah says, are correct. "It's the life-force energy that's giving you the answer," she says. I doubt that, but I do take pride in my educated-guessing abilities. That is, until another participant says the scuba diver has two children with another on the way, which certainly appears to contradict my answer. Yet Deborah declares that this too is correct. The diver, whom Deborah obviously knows, wants three children!

I assume Deborah is finding ways to make all of our answers right, and I figure she is doing this to boost our confidence so we'll keep working on becoming more clairvoyant. But if you start assuming you've gotten things right when you haven't, then how do you know what's real and what isn't? And what's the difference between "opening yourself up" to something that's genuinely out there and slowly talking yourself into a belief system?

Feeling thoroughly unswayed by the class and wondering what the next step on my mission to encounter healing energy should be, I hear about a chiropractor/energy healer named Donny Epstein who works with the motivational guru Tony Robbins. A friend of mine had gone to one of Robbins's advanced seminars and witnessed some of his fellow participants shaking in alarming ways while Epstein was "running energy." I also look into a Chinese qigong healer who participated in studies allegedly showing that he can kill cancer cells by sending qi energy to them. But first, I'm pulled in by a conversation with a woman from Texas named Carol Lee.

"Most everybody that comes to my classes sees light beings," she says from her car on the highway outside her home in Burleson, Texas, a suburb of Fort Worth. "It's because I tend to hold a pretty high vibration." Such "light beings," she says, come bearing some combination of wisdom, healing, love, and enlightenment.

"Like friendly ghosts?" I ask.

"More like angels."

Carol, who, like Deborah, also practices Quantum-Touch, says that if I come to one of her classes there is "no reason" I wouldn't see such remarkable things. "Every single person on this earth has this ability if you're open to it."

In addition to teaching Quantum-Touch classes and working with people individually, Carol also appears in Fort Worth twice a month for a holistic fair that's held at a place called Billy Bob's Texas, which bills itself as the "world's largest honky tonk." I find the idea of an energy healer doing chakra readings amid line dancing, country music, bull riding, and BBQ ribs to be oddly irresistible. But since I want to go to a class and Carol doesn't do those at Billy Bob's, I sign myself up for a weekend Quantum-Touch Level II class held in India-napolis, where Carol has a collection of devoted students.

It's a cool summer morning when I make my way to a bright and airy loft tucked into the back side of a Lilliputian office park. At the advice of Heidi, the Indianapolis-based Quantum-Touch practitioner who's coordinating the weekend, I've dressed in sheddable layers. "The temperature in the room is expected to rise and fall throughout the day as we run energy in the space," she informed me in an email. I envision energy like oil gushing from a busted pipeline, so plentiful we might not know what to do with it all.

When Carol greets me and the other seven women who have signed up for the class, I see that she does not have the joyful, serene presence I have come to associate with energy healers. Even as she instructs us to breathe in energy and imagine it filling up our hearts—to "let love breathe you"—she has a defiant, don't-tread-on-me Texas attitude. Carol declares most healers "far too spacey," always meditating and poking their heads in the clouds. Her own daily practice of meditation, she says, consists of "doing her hair and makeup." *Hogwash* appears to be one of her favorite words.

I'm disappointed to find not much talk of the hovering orbs of light that have lured me here. Instead, Carol does other demonstrations to impress nonbelievers such as myself. Quantum-Touch founder Richard Gordon insists that he can realign people's joints, hips especially, without even touching them, and he encourages everyone who learns Quantum-Touch to try this. Today, we are going to focus on moving the occipital bones, located at the back of our heads, at the threshold of the neck, without using any type of physical pressure. I am selected to go first, and a woman named Miranda puts her hands at my head to measure, declaring that my right bone is a bit higher than my left. Carol comes over to confirm. She is wearing a long black dress that hovers, spirit-like, over the floor. Every time she moves, four beaded necklaces clang together. "Oh, yes, it's definitely off to the right," she concurs, adding that such a thing is normal. "Most people have some misalignment, which is why this is such a good exercise."

Doctors will say that the bones of the skull don't move once they've stopped growing, so I am taken aback that the right side of my head does seem decidedly higher. Could my impression have been influenced by Carol's and Miranda's? It is, after all, a highly subjective measurement. Carol instructs everyone in the room to train their eyes on me and send their "heart energy" into my occipital bone. "The energy will go wherever you are focusing," she says. "You are bringing your love to the area." I feel the weight of fourteen eyeballs bearing down on me, and after a few minutes, Miranda remeasures. She declares that, lo and behold, my head bones have moved back into alignment. I reach back to feel, but this time it's hard to tell. Are they really more level? If anything, maybe the left is higher.

Carol doesn't wait for my confirmation. "Pretty amazing, right? You can do so much more than you realize."

The only abnormal experience I have during the weekend is a woozy sensation that comes on during a rapid-breathing exercise called fire breath. It starts in my legs and ascends to my torso and arms and eventually spreads all over. I start to feel tingly and weightless and decide that maybe I should keep at it to see where this goes. But then I suddenly realize I know this feeling. Life-force energy isn't filling me up—I'm hyperventilating. I had a similar sensation several times while I was pregnant, and I nearly passed out from low blood pressure. This militant breathing has lowered my carbon dioxide levels so that not enough blood is getting to my brain. I stop and do slow, careful inhales. The whole thing melts away.

That's the extent of my supernatural experiences. But on Sunday morning Carol leaves me with one last shot. Almost in passing, she mentions that she once "dropped a large man to the ground" using energy. After class ends, I ask her about this. If such things are possible, I say, I'm eager to see them. Could she demonstrate it to me? Carol says she can't do that. "It's not how the principles of the universe work. You don't just do it for fun or because someone wants you to. There has to be a higher purpose."

I suggest that rattling the worldview of a skeptical journalist might be such a purpose, but Carol doesn't buy it. "You have to do your own work and find your own truth."

"I was just hoping for something to nudge me along."

"Be careful what you wish for." Instead of walking back her boisterous claim she tells me a story of someone who, like me, wished to see spirits and auras, only to find herself one night being lifted from her bed. Carol also says she could make my water bottle fly across the room and spoons bend if she wanted to.

So instead of seeing a light being, witnessing bones moving, or watching beverages sail off tables, this is how it ends for me in Indianapolis. Still, I try not to harden around my skepticism. As I meet healers and get different treatments, I am ready for whatever might come. But nothing unusual happens. The most notable experience is a Reiki session I get while at the Cleveland Clinic's Center for Integrative & Lifestyle Medicine. There, in a relaxed and hazy stupor, I feel my Reiki healer's hands move even when she's not touching me, as though I were brushing up against something that isn't there.

Afterward, I tell her about the ghostly feeling, and she says, "I'm glad you could feel the energy." I don't mention that since I got the sensations only on my bare arms and feet, not from my clothed legs and torso, I was probably sensing body heat from her hands. She firmly believes it was energy. Like Carol, Deborah, and many others, she has her reasons for this conviction, some of which originated with her own visceral perceptions of such healing energy. But what if such seemingly transcendental occurrences are the result of something else?

In the spring of 2015, Al Powers, an assistant professor of psychiatry at Yale University, wandered tentatively out of his car and into a Best Western in North Haven, Connecticut, located just off Interstate 91. Inside the lobby, he stopped at a table where a man was seated next to a sign that read CONNECTICUT PSYCHICS ASSOCIATION.

The man looked at Powers. "Oh, you're the guy I talked to on the phone. What did you want to do again?"

Powers explained that he was hoping to learn about psychics for an upcoming research project. Did any people in the group have "auditory experiences other people don't have"? He didn't want to use the word *hallucinations*. "You know, like the experience you get with different forms of psychosis, but these people don't have any mental illness?"

"You mean people who are clairaudient," said the man, named Michael.*

"Sure, yeah, clairaudience."

This was the Connecticut Psychic Association's monthly gathering, where anyone could come and get a psychic reading from one of a dozen people holding court in one of the hotel's meeting rooms. Powers waited until a woman named Barbara* was finished with a client, and then Michael introduced them. Barbara was middle-aged

---

*Not their real names.

and worked by day as a police dispatcher but did psychic readings on evenings and weekends. She explained to Powers that since the age of six, she'd heard voices talking to her as clearly and audibly as if they were coming from a person seated next to her. They came nearly every day and told her stories, gave her information about people she knew, and informed her about what was going to happen next, none of which she found disturbing. She referred to several of the voices as her "spirit guides."

"Do you ever get angry or threatening voices?" Powers wanted to know. Barbara said that from time to time she did, but that over the years she'd devised ways to tune them out. Then he asked her if she would be willing to come into his psychosis research lab at Yale for testing. He assured her he didn't secretly think she was psychotic, but was interested in studying people like her to understand if her experiences could shed light on new ways to treat people with schizophrenia who also hear voices.

Inside the Yale lab, located in a locked psychiatric unit, Barbara was asked a long battery of detailed questions that had been formulated by forensic psychiatrists to determine whether a person's experience of hearing voices was genuine. She was asked about the gender and age of the voices, whether animals ever spoke, where in space the voices seemed to come from, and what their syntax was— past or present tense, first or third person. Over hundreds of such interviews, psychiatrists have determined that certain phenomena are common and others rare. Hearing only the voices of children, voices that change gender midsentence, and voices that are almost always inaudible or screaming are, for example, rare. If a person claims a high number of these experiences, it's a sign they're probably not being entirely truthful. Powers was pleased to see that Barbara passed with flying colors. She was fully functioning and had an important job—police dispatcher—and just happened to hear voices in her head all the time.

Powers and his Yale colleague Philip Corlett tested and confirmed sixteen more such "clairaudient psychics" and included them in a

2017 study that compared their experiences to those of voice-hearing participants with schizophrenia. The study found that while the voices of both groups had similarities—their frequency and content, for instance—people interpreted them with revealing differences. "Instead of trying to block the experience, the psychics embraced it and interacted with it. They see it as a gift rather than a curse," says Corlett. The psychics said they used their voices to try to help people, such as during psychic readings, when the voices, or "spirit guides," connected with a client's deceased relatives or loved ones. The participants with schizophrenia, on the other hand, were troubled by and sometimes frightened of their voices. Unlike the psychics, they did not think they had any control over them or that they served any useful purpose.

The study didn't look directly at why some people with voices were diagnosed with schizophrenia and needed treatment while others used metaphysical interpretations to lead happy, functional lives, but there was one telling detail. The average age the psychics reported first hearing their voices was seven, whereas for the people with schizophrenia it was fifteen. "You could imagine the parental reaction to a seven-year-old saying they're hearing something being very different to that of a fifteen-year-old," Powers says. "A lot of people don't realize that if you have hallucinations, it doesn't necessarily mean you have a mental disorder."

Hallucinations are much more ordinary than people generally acknowledge. No one can say for sure how many people without mental illness have the semiregular experience of hearing, seeing, feeling, or smelling things that aren't there, but estimates put it at around 10 to 15 percent. Given the right circumstances, even greater numbers of us can experience hallucinations. In a 2009 study at University College in London, two-thirds of volunteers who scored high on a hallucination proneness scale went into a sensory-deprivation chamber and started experiencing unusual perceptions within as little as fifteen minutes. Some reported seeing faces and shapes hovering near them. Others heard strange noises or noted an unusually heightened sense of smell. Two sensed an "evil presence." This happened less frequently to the volunteers who were less hallucination-prone, but roughly 20 percent of them still reported bizarre and phantom sensations.

Researchers explain that these incidents occur because the brain, starved of sensory input, begins generating additional activity, which people then misattribute to an external source. It's like what happens to older people who get Charles Bonnet syndrome after they lose their vision. Although their eyes have degenerated, their brain's visual cortex is still raring to go, giving these blind elders the perception of "seeing" faces, animals, pieces of furniture, and other objects they used to perceive.

Hallucinations can also be emotionally driven. Between 30 and 50 percent of elderly widows experience what are called bereavement hallucinations, and more than two-thirds of them find the experiences pleasant or helpful, thanks to the emotional comfort they give and the lessening of feelings of isolation. After his parents died, the astrophysicist and skeptic Carl Sagan wrote about their visits in *The Demon-Haunted World*: "Probably a dozen times since their deaths, I've heard my mother or father, in a conversational tone of voice, call my name. . . . I still miss them so much that it doesn't seem strange at all that my brain will occasionally retrieve a lucid recollection of their voices."

Though not necessarily regarded as hallucinations, phantom bodily sensations can also be easily induced, as researchers demonstrated in a 2015 study in which 90 percent of volunteers felt moderate or strong sensations from nothing at all. When given a placebo skin irritant and a red laser that was not on, people reported sensations much like those I heard described during acupuncture or energy healing sessions, such as tingling, warmth, and "pulsing." Some participants said the feelings traveled along pathways in the body, from the little toe, where the fake irritant and red light were applied, to the thigh, and from the little finger to their upper arm and shoulder.

Derek Tracy, a research psychiatrist who runs an NHS-funded lab in London, tells me of an experience he had a few years ago. One night before going to sleep he opened his eyes and saw a dog in his bedroom. Startled, he remembered he didn't own a dog and sat up to get a good look at it. But the dog had already vanished. "Some people might have started to think, 'Well, I had a dog as a child and it looked a lot like that dog,' or 'My dog just died, what does this mean?' But I'm a scientist, so I think, 'Well, I'm in a quiescent state and my visual cortex has nothing else going on.' I disregard it and go back to sleep."

Tracy says he's also had experiences of feeling his phone vibrate in his pocket, only to find that it's in another room or tucked into a computer bag. "We all have what you could call hallucinatory experiences on a fairly regular basis," he says. "Our brains have evolved to make sense of incomplete information and to do this very rapidly, and it's part of our survival as a species. Our brains are always going, 'What's happening? What's happening? Join the dots, join the dots.' It's a very rapid, but error-prone system. But from an evolutionary point of view, you don't want the perfect slow system."

I describe some of the unusual experiences I've heard about—the bright white lights, the auras, the orbs, the tingly, warm bodily sensations, the stabbing hands, the disappearing legs—and Tracy says it's quite probable that these were forms of hallucinations. Such perceptual errors are most prone to occur in times of sensory deprivation, amid heightened expectations and suggestion, and during altered states of consciousness, such as those sometimes achieved while meditating or during the low-input, transitional period between wakefulness and sleep—all of which are often present during healing sessions. "Who goes to a healer or becomes a healer? You have a self-selecting cohort of people with certain beliefs and expectations, and so they're more likely to interpret their hallucinatory experiences in a certain way. I think it comes down to how you make sense of it and what you do with it." Much as the Connecticut psychics view their voices as messages from the beyond, many people in alternative medicine read their brain's perceptual quirks or shifted states of consciousness as manifestations of healing energy.

Or maybe it's not always quite that simple. Donna Eden, the healer I spent time with in Asheville, has nonstop anomalous experiences. Unless she actively works to blot them out, she sees perpetual patterns of color around people's bodies. That they don't come and go would seem to rule out the idea of hallucinations. Instead, her perceptions sound more like something I read about in a 2004 paper by Jamie Ward, a cognitive neuroscientist at the University of Sussex. In the early 2000s, Ward became interested in the fascinating phenomenon of synesthesia, which involves the cross-processing in normally distinct brain regions. In the most common forms of this condition, letters, numbers, days of the week, or months of the year appear in particular colors. A person might always see the word *Tuesday* in yellow, for example, or the number five as dark purple. As a child, the novelist and famous synesthete Vladimir Nabokov insisted that all his alphabet blocks were in the "wrong colors." The letter *B*, he would later explain, is burnt sienna. *M* is "a fold of pink flannel."

Other synesthetes perceive sounds as having color. To create his colorful abstract paintings, Wassily Kandinsky is said to have sometimes turned on music and painted what he saw. Or there can be a cross-processing of sounds and tastes. A few years ago, a British synesthete told a flabbergasted interviewer that the word *speak* has always tasted to him like bacon. He had no idea why. Estimated to occur in 1 to 4 percent of the population, synesthesia is constant throughout one's life and has a strong genetic component. Forty percent of synesthetes have a parent, sibling, or child with the condition, though not necessarily the same form of it.

Searching for research subjects, Jamie Ward met a woman, a student at the University of Sussex, who didn't quite fit into one of these more commonly known varieties. When he brought her into his lab, she told him that she saw colors around people she knew. Those she felt positively about appeared in yellow, pink, purple, and green tones, and those with whom she had negative associations bore black, red, gray, silver, and brown hues. If someone was angry, the person would appear red. Being at a fun party would tint the whole room pink. After verifying that her experiences were authentic by doing variations of the Stroop test,* Ward concluded that G.W., as he called her, had a new form of synesthesia in which emotions are mapped to colors. He dubbed it emotionally mediated synesthesia. Later, the neuroscientist V. S. Ramachandran at University of California, San Diego, referred to it as emotion-evoked colors. Others have called it person-color synesthesia.

In his paper, Ward noted G.W.'s obvious similarity to people who claim to be aura readers or "sensitives." "Although many people claiming to have such powers could be charlatans," he wrote, "it is also conceivable that others are born with a 'gift' of synaesthesia." G.W. does not believe she has any mystical powers nor does she have any interest in the occult, yet Ward said he could see how such interpretations might arise. "You can understand how people who have this and were born in a different age or who come to a certain set of beliefs would consider themselves able to see spiritual states."

With her life color readings and detailed impressions of people's emotional states, it's not hard to see how Donna Eden's brain might be engaged in a similar mapping of feelings to colors. Yet her experiences are also quite different: For one, they are far more elaborate, involving seven different aura layers and seven layers within each chakra located along the midline of the body. There is also movement to the colors, with some chakra layers spinning in a clockwise or counterclockwise direction. And while G.W. and other emotionally mediated synesthetes need to get to know someone before seeing their colors, Donna appears to be able to quickly tune in to someone's feelings, personality, and state of mind.

"Nobody's ever tested me on that," Donna says when I ask her about synesthesia. "But I do know there are certain tastes where I know the color of that taste." One night recently, she says, she was

---

*To give yourself a Stroop test, have someone write the names of colors in differently colored markers. For instance, write the word *red* with a blue marker, *yellow* in orange, and so on. Then try to name the color the word is written in and see if your brain stumbles.

out at a restaurant with her husband, David, and as a dessert floated by on a tray, he remarked, "God, that looks beautiful." But Donna didn't think so. To her, it had the taste of a putrid color, "like a rosy green." To her, the best-looking tastes are Dairy Queen sundaes, her all-time favorite food. "They're a very mellow, mellow yellow, almost white, sometimes with lavender going through it."

In addition to possibly having taste-color synesthesia, Donna may also have two other kinds: Kandinsky's sound-color variety and what's called mirror-touch synesthesia, where people spontaneously feel the emotions and bodily sensations of others. Donna says that from time to time she has been overcome by other people's symptoms, and she once became temporarily paralyzed while working on someone who had French polio or Guillain-Barré syndrome. Over time, she says, she has figured out how to prevent or quell such occurrences.

Donna doesn't think synesthesia fully explains her experiences. She believes she is using her visual system to pick up a different and unordinary aspect of the physical universe—the subtle energies that animate the body. Certainly, without testing we can't say for sure whether she has one of the known forms of synesthesia, never mind whatever emotionally mediated variant might help explain her kaleidoscopic visions. But it does offer a possible explanation for claims that seem both extravagant and wholeheartedly honest. I think Donna's visions are as real as the voices heard by the Yale psychics. It's the interpretations she gives them that are hard to take at face value.

Nor do I think that she and other alternative healers are actually testing someone's energy levels when they do those muscle tests with people's arms—those demonstrations I saw Donna do onstage in Asheville and that blinded studies have failed to validate. Yet something is happening to make a person's outstretched arm withstand downward pressure one minute and then cave the next. While trying to understand this, I come across a posting on a chat board by a chiropractor named Stephen Perle, who says he used to be an enthusiastic muscle tester but now has a very different opinion on the subject. Perle is a professor of clinical sciences at the University of Bridgeport's chiropractic college. When I call, he tells me that he now believes that during the several years he was doing muscle testing (also called applied kinesiology), in chiropractic school and immediately after, he was unconsciously pushing in a different way when he thought someone was going to be strong than when he thought the person was going to be weak, or when he thought the question he was asking would get a yes rather than a no. When a person seemed strong, he says, he would contract his arm and shoulder muscles in a manner that was never going to move anything. These isometric contractions can look as if they involve impressive amounts of force, but they're really stationary maneuvers. Only when Perle intuitively felt that his question or action would cause someone's arm to test weak did he engage his muscles to actually create movement and get the arm down. Perle says that subtle manipulations in timing and pushing at the far end of someone's extended arm, as opposed to closer to the shoulder, help give the pusher a considerable advantage. Unless someone is exceptionally strong or has sprinter reaction times, it's relatively easy to get his or her arm down.

But just as he wasn't doing this intentionally or trying to pull the wool over anyone's eyes, Perle doesn't think most people who do muscle testing are conscious of what's happening either. Nor does he think the person getting tested participates in any way: "All the force variation comes from the person doing the testing. I think they're externalizing something that they know or have decided internally." Much like hallucinations, the information coming to them is being generated from within. It just feels external. "I really don't think it's a conscious attempt. They're being honest and self-deceived at the same time."

Why Doctors Need to Be More Like Alternative Healers (and Vice Versa)

In the spring of 2017, a few months before heading off to Tübingen, Germany, I got a call from my dad's wife, Kathy. In a hushed, frightened tone, she told me he was in the ICU and not doing well.

Since I had been in only sporadic contact with my dad, I'd always feared getting a call like this. In the handful of conversations we would have each year, he never revealed much about his health. I knew he was generally happy with his life, but I had no idea how he was doing physically, and as he was a man in his seventies, I suspected there were issues. But none had been revealed until now.

Kathy and I spoke for a few minutes and then she passed the phone to the doctor who was in the room. "Your dad might not make it through the night," he warned. Upset and agitated, I packed a small bag and got onto a plane that landed in Oklahoma City the next morning.

When I arrived, my dad was in a coma and lying rigidly on the bed. He was attached to so many tubes I couldn't tell where one ended and another began. There was a breathing tube, a catheter, a dialysis machine, and a variety of IV drips, all working to keep him alive and reverse the toxic shock his body had gone into. My dad had many more underlying health problems than even Kathy had realized. He had stage two kidney disease, kidney stones, and pneumonia, which he had apparently been silently harboring for some time. When he finally decided to acknowledge that something was wrong and asked Kathy to take him to the hospital, it was a perfect storm from which there could be no return. His kidneys and lungs weren't responding to the help they were getting from the dialysis machine and the breathing tube, and his oxygen levels began dipping dangerously low. The doctors suspected he already had significant brain damage. So within twenty-four hours of my arrival, Kathy and I allowed the nurses to remove the tubes, and ten minutes later, he died.

Watching my dad go felt far more profoundly sad than I'd thought it would. But as overcome as I felt, I knew this was going to be far harder on Kathy. She and my dad had been together for the better part of two decades, and in recent years, after retiring and moving to Oklahoma from the East Coast, they had been inseparable. "He was the best husband in the world to me," she said, not long before he left her forever. He was a source of love and companionship, and a rock of unremitting help and support. For nearly a decade, Kathy had been struggling with throat cancer, and my dad had devoted himself to her care. It was always her health they talked about, which is maybe why he'd let his own slip away. Her cancer had been in remission for a few years, but it had recently spread to her lungs and she had been making plans for a new round of chemotherapy. Then there was the fibromyalgia. Kathy had continuous pain in her knees and back and sometimes suffered achiness all over her body. Just a few weeks ago, she had also gotten back surgery after taking a bad fall. Now she would be facing all this without the shoulder she had come to lean on.

I wasn't sure what to do next or how long I should stay in Oklahoma. Kathy wasn't going to do a memorial service right away, and she and I didn't know each other that well. But we ended up spending quality hours together back at the hospital the next morning. During the night, Kathy had woken up seized with a sudden feeling of pain in her chest and a sense that she was having a heart attack. She called 911 and was taken, as a cruel irony would have it, to the same ER room my dad had been in before being transferred to the ICU. When doctors examined Kathy, they determined that she wasn't having a heart attack. But just to be sure, they ran a number of cardiac tests and took several vials of blood.

Not long after I arrived in Kathy's room, a young, handsome cardiologist came in to deliver the results. He told us what we both already knew: her circulatory system was fine. The pain in her chest and the trouble breathing had been a panic attack, brought on, no doubt, by the events of the past three days. Standing at the foot of the bed, the cardiologist told Kathy she could rest assured. Everything was fine and she could get ready to go home soon. I suppose it should have been a relief for Kathy to hear her heart wasn't failing, but she didn't look especially relieved. I wondered what she was supposed to do next. She still had cancer and fibromyalgia and my dad was still gone. "What happens if she starts getting chest pains again?" I asked the cardiologist. "Is there anything she can do?"

At this, the cardiologist looked uncomfortably at his feet and said, "I wish I knew. I think if someone could figure that out, they should write a book about it."

In fact, people have written books—with titles such as *When Panic Attacks* and *Overcoming Panic Attacks*—but I hadn't meant this as a literary challenge. I more envisioned a handout on calming breathing techniques or a referral to a website or someone Kathy could talk to if she was so inclined. The ideal thing would have been a cognitive behavioral therapy class tailored to panic attacks, but such things are few and far between. It didn't seem right to simply ship her off to an empty house.

I'm sure some doctors would have known exactly what to do and say in this situation, but this obviously wasn't one of them, and the experience left me with a deep understanding of why alternative medicine exists. I have no doubt that many of the acupuncturists, energy healers, chiropractors, meditation teachers, and tai chi instructors I met while writing this book would have had something useful to offer a patient like Kathy. I imagine them, instead of standing nervously at the edge of the bed, going over to sit down next to her and gently acknowledging the torrent of feelings that were obviously overwhelming her. Maybe they would have given her something to think about or to visualize or some points to press on the next time she felt an attack happening—some small but not insignificant measures of armor against another fruitless trip to the ER.

Since embarking on this journey, I've learned that this is what alternative medicine does for people. By making us feel supported, by summoning the power of expectations and belief, by relaxing our bodies and reducing stress, mind-body therapies move molecules in our brains in a way that can reduce the ills we feel in our bodies. They shift those symptoms for which brain activity plays a significant role—pain, panic attacks, fatigue, shortness of breath, nausea, psychosomatic neurological problems, and other unexplained symptoms—even though we're far from a complete scientific understanding of how exactly they do all this. These therapies can also calm, empower, and inspire us, giving us more resilience than we imagine we have, and leading us down paths we wouldn't otherwise find. They do things that standard medicine doesn't often pay enough attention to or quite know how to handle.

But there are limits to what alternative medicine can do, and knowing this, I think, is a crucial part of deciding when to seek it. Although some of these therapies have clear physical impacts, such as lowering the stress response, reducing inflammation, easing muscle tension, and occasionally helping certain neurological problems, as a general rule they do not eradicate physical disease or directly repair substantial damage to tissues.

As Harvard's Arthur Kleinman noted back in the 1970s, only medical doctors are theoretically capable of treating both aspects of human health. There's no reason why a cardiologist can't be a perceptive healer who supports a patient after her grief-induced panic attack. Or why an oncologist can't administer a cancer treatment while helping her patient cope with its brutal side effects. Or why a rheumatologist couldn't prescribe drugs to rheumatoid arthritis patients while also helping them better manage the life stresses that

can trigger inflammatory flare-ups. Yet when I spoke with Kleinman, he told me that, despite what are often good intentions, the
culture among practicing doctors is still not conducive to forging
these empathetic connections with patients or treating their illnesses
as well as their diseases, partly as a result of the "devastating effects
of medicine as big business." During medical school and residency
training, the culture among doctors is even worse. "Toxic," he called
it. In a small but revealing 2004 study, researchers at the University
of Liverpool reviewed audiotaped consultations between twenty-
one general practitioners and thirty-six patients who had medically
unexplained symptoms. All but two of these patients gave doctors an
opening to discuss social or emotional difficulties they were having.
Many spoke of their stress levels or moods. But most GPs missed the
opportunity. They "did not engage with these cues," the researchers
wrote. Another study found that even though 60 to 80 percent of
primary care visits in the US are likely to have a stress-related component, only 3 percent of visits include discussions of stress management. I wondered why.

Even after three years on the job, Dr. Edward Hundert still gets
excited by the pods. These are the rooms, eighteen in all, outfitted to
look just like a doctor's office. They have an examination table with
the crinkly sheath of paper, counters and cabinets of supplies, a laptop
stand, a wheeled stool, and anatomy charts pasted on the wall. They
are also rigged with surveillance equipment so that when Harvard
Medical School students go into them to talk to a paid actor who is
pretending to be sick, professors can sit on the other side of a one-
way mirror and see how well America's future doctors are handling
the job. Later, these professors and the students drill down for closer
inspection with video recordings. "We have two of them in there,"
Hundert says, motioning to the small cameras mounted on the ceiling, "so you can get a really good look at both the patient's face and
the student's as they're talking to each other."

As the dean for medical education at Harvard Medical School, Hundert considers it his job to train students to be just as skilled at communicating with patients as they are at diagnosing specific health problems—to meet Kleinman's standard of treating both illness and disease. The pods play a big role in this. Inside them, a whole menagerie of patients appear: a retiree who is belligerent or difficult, a young mom who isn't admitting to using drugs, an athlete with a torn ACL whose dad was just diagnosed with cancer, a man with a conversion disorder who won't make eye contact. Students must figure out what tests should be ordered and what treatment the patient might need, and they are expected to go about this in a manner that creates trust and makes the patient feel understood, which Hundert says is usually the far harder part of the equation.

"What happens is that the first-year students do a decent job of talking with patients, but as they go on and learn more knowledge, the pressure to get answers to all these specific questions starts to intrude." Students forget about sitting down at eye level with the patient and spend too much time tapping away at their laptop taking notes. They also forget to ask the open-ended questions that often get patients to open up. For instance, when patients report that they've been saddled with recent episodes of depression and abnormal levels of fatigue, the student doctor might ask, "When did this start? Is it worse in the morning or later in the day?" What the doctor should say is "Tell me more about that."

"There's a lot of psychiatry in all of medicine. So much of what's coming in to hospitals and doctors' offices is not what it seems. We have this joke: 'What's the difference between an ER doctor and a psychiatrist? A psychiatrist knew they were getting into psychiatry.'" Hundert even ventures so far as to say that all this psychiatry, once relegated to a "good bedside manner," is more important than technical skills and knowledge.

I've come to Harvard, the nation's top-ranked medical school and Kleinman's home for the past four decades, to find out why there aren't more doctors who know how to understand and treat patients' minds as well as their bodies. Before going to Harvard Medical School himself, Hundert studied philosophy at Oxford, and it shows. He is erudite and thoughtful, though not to the point of severity. His near-constant smile renders him a highly approachable presence on campus.

I'm getting the impression that medicine's failure to generate more empathy among doctors is not for lack of trying, at least not at Harvard. As we stride away from the pods toward Hundert's office in the adjoining building, he explains some of the changes Harvard made four years ago to its curriculum. Instead of doing intervals of clinical work at nearby hospitals in their second and third years, students now interact with real patients almost from day one. Every Wednesday, they leave the classroom to join a primary care doctor in his or her office. The idea behind the move, Hundert says, is to provide a two-year continuous real-world context for the volumes of knowledge students are learning about disease and bodily systems. It's also a means of hammering home that practicing medicine isn't just about trying to mend the body; it calls for engagement with the messy, complex emotional life of the person inhabiting it.

Harvard does more to impress this upon its students. In Practice of Medicine classes, they are taught that, in addition to all the necessary tests, measurements, and cataloging of symptoms, they must get a patient's social history—details of their family situation, their job stressors, financial troubles, and anything else of importance in their lives. Professors have students write "illness narratives," in which they tell a story from a patient's point of view, and read a famous 1927 essay by the Harvard professor Francis Weld Peabody, in which he writes, "One of the essential qualities of the clinician is interest in humanity, for the secret of the care of the patient is in caring for the patient." Students are even told about the value of touching patients, perhaps placing a hand on their hand or holding a shoulder while checking their lungs.

"You're trying to symbolically sit next to the patient and look out together at this problem that they have. 'How are we together going to attack this thing?' When you have that alignment, you can get a lot of healing done." Hundert doesn't like to use the word *healer*, even though this is what he's talking about, because he worries that when used to refer to doctors, it suggests a paternalistic dynamic in which the doctor holds all the power.

I ask Hundert whether it's possible for today's doctors to dig beneath the surface like this in the fifteen to twenty minutes they are allotted for each of the eighteen to twenty patients who cycle through on a given day. "Yes, it is exhausting," he admits. "If you had more time, you could do it better, and if you didn't have the electronic-medical-record constraint, you could also do it better, because you're looking at the screen, not the patient. But I'm convinced you can do more in a short amount of time than people think."

He says that back in the 1980s, during his internship at Mount Auburn Hospital in Cambridge, he had a mentor who was among the hospital's busiest doctors. "We interns would go into the rooms and the patients would say, 'Oh, I was talking with Dr. Grabowski this morning, and he said this and he said that.' It always sounded like they had had a long conversation with him." But Hundert knew the bustling Dr. Grabowski was spending only a few minutes with each patient. Hundert wanted to figure out what was going on and asked if he could follow Dr. Grabowski around. Hundert discovered that Dr. Grabowski was sitting down with patients in each room and asking them open-ended questions like "What's on your mind?" "How are you doing?" "What do I need to know?" He never said, "How's the knee?" "Immediately, the patient got onto what they thought the doctor needed to hear. Dr. Grabowski gave the patient such high-quality attention that they really felt like they had been heard. The amount of time they spent with him felt adequate."

Despite attempts here and at other medical schools to teach doctors this kind of empathy, it often doesn't stick. "I think we have a very long way to go if you look at the patient satisfaction scores," Hundert says, referring to the Centers for Medicare & Medicaid Services' most recent survey, in which only 40 percent of US hospitals received a four- or five-star rating out of five for patient satisfaction. Other polls show that only 34 percent of Americans have great confidence in the leaders of the medical profession. Even more distressing are studies that evaluate medical students before and after med school, as well as before and after their residency. Results show that empathy scores decrease as people move through the system. Just as with the pod tests, the greater the volume and complexity of medical knowledge, the greater the tendency for the squishy stuff to get brushed aside. Understandably so. Unlike alternative healers, doctors are responsible for making sure patients don't die from serious problems or aren't misdiagnosed when diseases need to be quickly treated. Failure to order the right test or look closely enough at the results can be catastrophic.

We walk across the grassy quad that strings together Harvard's stately med school buildings. Hundert runs into a first-year student to whom he owes an email. There's a quick hello and a promise to connect tomorrow, then Hundert turns to me and says, "It's an amazing story. He's from Gambia and is the first person in his family to go to college, never mind med school." Hundert pauses, still beaming from the thought. "I think helping people is the primary motivator that draws people to medical school. Maybe this didn't used to be the case ten or twenty years ago when doctors made more money or could be more of their own boss. But now I see that very pure desire. The trick is not to beat it out of them."

In Asheville, Donna Eden said, "We heal one another just by recognizing each other," and in Boulder, Peter Churchill spoke of "primal human sharing." If they're skilled at what they do, alternative healers already know all of the things Hundert is trying to teach at Harvard. Since they don't prescribe drugs or perform surgeries or have to worry about monitoring organs or isolating infections, energy healers, acupuncturists, chiropractors, and all the rest have intuitively figured out that shifting a person's inner ecosystem can

work far bigger wonders than most of us realize. Mind-based healing, whether you call it a placebo effect or something else, is real and
can in the right conditions be life changing, even though scientists
have scratched only the surface of the mechanisms. But regardless of
how such healing works, most of the people I've written about in this
book know exactly how to elicit it. Doctors could learn a thing or
two from them. "We physicians must be humble before our ignorance
of how one person can relieve the suffering of another," wrote the
Yale gastroenterologist Howard Spiro.

But the door needs to swing the other way, too. Back in San Diego
at the Pain Summit, I had breakfast with a woman who called herself a "walking placebo." A Chicago-based physical therapist, she had
years earlier realized that the nature of the effect she was having on
her patients with chronic pain was very different from what she'd
thought. "What you're doing is interacting with another human
being. You don't have to layer on a lot of crap," she said, referring to
all the fancy techniques she had learned. She was the only full-time
practitioner I met who would cop to such an honest acknowledgment
of what was going on. Everyone else was a believer in his or her own
placebo. Although they embraced psychological aspects of their work,
they also clung to the other stuff—the qi, the quantum and life-force
energy fields, the meridians, the chakras, the disruptions in "etheric
bodies," the realigned vertebrae, the subluxations—all concepts that
undermine credibility and help keep alternative medicine alternative.

Although it's tempting to say these ideas should just be taken as
metaphors—placeholders for biological concepts we don't yet have
the words for—the fact is they're firmly baked into the cake. How can
you teach someone acupuncture without meridians or chiropractic
without explaining all the different vertebrae to realign? How would
people know where to put the needles or push with their hands?
Who would pay to go to school to learn a system of allegories? Plus,
believing in something inspires a strain of confidence that can end up
infecting patients. Scientists don't know enough yet to offer better
guidelines.

I don't have any terrifically satisfying answers for these questions because, while I'll always side with science and couldn't possibly defend the flood of nonsense often gushing forth from the alternative medicine community, I've also seen these therapies help people, and the stories they weave provide people with a large measure of meaning and comfort. So until more doctors figure out how to do the things alternative healers do, to treat the patient and not just the disease and to evoke the power of belief and hope; or until solutions such as mindfulness-based stress reduction and cognitive behavioral therapy tailored to specific problems become ubiquitous in medicine; or until the Germans bring their holistic psychosomatic departments over here, the point is moot. Alternative medicine isn't going anywhere as long as it continues to give people something they're desperately missing.

Acknowledgments

I owe so much to the people who helped me write this book. First and foremost are all those who generously shared their stories of personal healing and transformation. I was continually humbled by their accounts of suffering and awed by their resilience. Although many of these stories do not appear in the book, they are among my original sources of inspiration. I am also deeply indebted to the dozens of researchers who offered their ideas and helped me unravel the complexities of brain, placebo, and pain science, especially those I pestered with multiple lines of inquiry and who patiently endured my seemingly eternal confusion. For rousing criticism of alternative medicine and an unwavering adherence to critical thinking and knowable facts, I am grateful for the skeptics. Although I think they have too little interest in the mind's role in health and healing, in the many hours I spent reading and digesting their arguments, I found them to be right far more often than they were wrong.

It's impossible to say too many positive things about my experience working with Scribner, both for this book and my previous one. I am lucky to have an editor as thoughtful, diligent, and encouraging as Daniel Loedel and to be the beneficiary of his nimble touch on the page. Nan Graham has been an enduring source of support and a saint for her patience with missed deadlines.

None of this would have happened, however, without my exceptional agent, Molly Friedrich, who first suggested that I go off and write a book about alternative medicine. For their input and words of wisdom, much gratitude goes to Lisa Ferri, Laura Rich, Mariem Horchani, Hillary Rosner, Florence Williams, Rachel Horner, Kerry Nankivell, Jane Chun, Karen Leo, Harlan and Ingrid Meyer, and especially Hannah Nordhaus and Stacy Griffin, who suffered through dreadful first drafts. Hannah's gift for editing is rivaled only by her writing abilities, and she made me look far smarter than I am.

Finally, there are my two wonderful sons, Jude and Luke, who give me daily joy, and my husband, Rich, whose support and insight never fail to get me through any problem I bring before him. I am blessed to have you in my life.

Notes

Chapter One: Donna's Eden

4   *people she has trained as practitioners:* http://www.innersource.net/em /resources/case-histories.html.

5   *even cancer, particularly of the kidney:* G. B. Challis and H. J. Stam, "The Spontaneous Regression of Cancer: A Review of Cases from 1900 to 1987," *Acta Oncologica* 29, no. 5 (1990): 545–550, https://www.tandfonline.com/doi /pdf/10.3109/02841869009090048.

5   *diagnostic errors happen to one in twenty Americans:* Hardeep Singh et al., "The Frequency of Diagnostic Errors in Outpatient Care: Estimations from Three Large Observational Studies Involving U.S. Adult Populations," *BMJ Quality & Safety* 23, no. 9 (September 2014): 727–31, https://www.ncbi.nlm.nih.gov/pmc/articles/PMC4145460/.

5   *"Until very recently, the history of medical treatment":* Arthur K. Shapiro and Elaine Shapiro, *The Powerful Placebo* (Baltimore: Johns Hopkins University Press, 1997), 11.

6   *Once called the "coolest strangest thing in medicine":* Ben Goldacre, "All Bow Before the Might of the Placebo Effect, It Is the Coolest Strangest Thing in Medicine," *Bad Science*, March 1, 2008, https://www.badscience.net /2008/03/all-bow-before-the-might-of-the-placebo-effect-it-is-the -coolest-strangest-thing-in-medicine/.

6   *In any given year, across the US:* "Use of Complementary Health Approaches in the U.S.," National Center for Complementary and Integrative Health (NCCIH), https://nccih.nih.gov/research/statistics/NHIS/2012/mind -body/manipulation.

6   *"by far the most successful claim to knowledge accessible to humans":* Carl Sagan, *The Demon-Haunted World* (New York: Ballantine Books, 1996), 28.

12   *Serbian tennis star Novak Djokovic says:* Novak Djokovic, *Serve to Win* (New York: Zinc, 2013), 22–23.

12 *In one of the more recent examples:* Stephan A. Schwartz et al., "A Double-Blind, Randomized Study to Assess the Validity of Applied Kinesiology (AK) as a Diagnostic Tool and as a Nonlocal Proximity Effect," *Explore (NY)* 10, no. 2 (March–April 2014): 99–108.

14 *William Osler, one of the founders of modern medicine, wrote in 1904:* William Osler, *Aequanimitas* (Philadelphia: P. Blakiston's Son, 1904), 273, https://books.google.com/books?id=VSYJAAAAIAAJ&printsec=frontcover&source=gbs_ge_summary_r&cad=0#v=onepage&q&f=false.

Chapter Two: Lost in Translation

23 *Voltaire advised young men to find wives:* Shapiro and Shapiro, *Powerful Placebo,* 13.

23 *doctors took eighty ounces of blood:* David M. Morens, "Death of a President," *New England Journal of Medicine* 34, no. 24 (December 9, 1999): 1845–49, https://static1.squarespace.com/static/54694fa6e4b0eaec4530f99d/t/56366137e4b0e32e6d545a7f/1446404407774/Death+of+Washington+NEJM.pdf.

23 *a similar revolution in human thinking was going on in ancient China:* Paul U. Unschuld, *Medicine in China* (Berkeley, Los Angeles, and London: University of California Press, 1985); Paul Unschuld, *What Is Medicine?* (Berkeley, Los Angeles, and London: University of California Press, 2009).

26 *Daoguang emperor issued an imperial edict:* Hanjo Lehmann, "Acupuncture in Ancient China: How Important Was It Really?," *Journal of Integrative Medicine* 11, no. 1 (January 2013): 45–53, http://www.jcimjournal.com/jim/FullText2.aspx?articleID=jintegrmed2013008.

27 *According to research done by Hanjo Lehmann:* Hanjo Lehmann, *"Akupunktur im Westen: Am Anfang war ein Scharlatan"* [Acupuncture in the West: In the Beginning Was a Charlatan], *Deutsches Ärzteblatt* 107, no. 30 (2010): A-1454 / B-1288 / C-1268, https://www.aerzteblatt.de/archiv/77695/Akupunktur-im-Westen-Am-Anfang-war-ein-Scharlatan.

Chapter Three: Telltale Toothpicks

30 *the law . . . mandated insurance coverage for any provider licensed:* "Every Category of Healthcare Providers, WAC 284-170-270," Washington State Legislature, http://apps.leg.wa.gov/wac/default.aspx?cite=284-170-270.

31 *Cherkin decided to do a large, well-designed acupuncture trial:* Daniel C. Cherkin et al., "A Randomized Trial Comparing Acupuncture, Simulated Acupuncture, and Usual Care for Chronic Low Back Pain," *Archives of Internal Medicine* 169, no. 9 (May 11, 2009): 858–66, http://www.ncbi.nlm.nih.gov/pmc/articles/PMC2832641/.

31 *this condition was—and still is—both the most common type of chronic pain:* "Diseases/Conditions for Which CAM Is Most Frequently Used Among

Adults—for 2007 and 2002," NCCIH, https://nccih.nih.gov/research/statistics/2007/diseases-conditions-for-which-cam-is-frequently-used-among-adults; Anna Shmagel et al., "Epidemiology of Chronic Low Back Pain in U.S. Adults: National Health and Nutrition Examination Survey 2009–2010," *Arthritis Care & Research* 68, no. 11 (November 2016): 1688–94, https://www.ncbi.nlm.nih.gov/pmc/articles/PMC5027174/.

31 *Studies have shown that between 20 and 40 percent of those going under the knife:* Zafeer Baber and Michael A. Erdek, "Failed Back Surgery Syndrome: Current Perspectives," *Journal of Pain Research* 9 (2016): 979–87, https://www.ncbi.nlm.nih.gov/pmc/articles/PMC5106227/.

33 *In other large acupuncture trials for headaches, knee osteoarthritis, and back pain:* Michael Haake et al., "German Acupuncture Trials (GERAC) for Chronic Low Back Pain: Randomized, Multicenter, Blinded, Parallel-Group Trial with 3 Groups," *Archives of Internal Medicine* 167, no. 17 (September 24, 2007): 1892–98; Benno Brinkhaus et al., "Acupuncture in Patients with Chronic Low Back Pain: A Randomized Controlled Trial," *Archives of Internal Medicine* 166, no. 4 (2006): 450–57; Hanns-Peter Scharf et al., "Acupuncture and Knee Osteoarthritis: A Three-Armed Randomized Trial," *Annals of Internal Medicine* 145, no. 1 (2006): 12–20; Claudia Witt et al., "Acupuncture in Patients with Osteoarthritis of the Knee: A Randomised Trial," *Lancet* 366, no. 9480 (July 9–15, 2005): 136–43; Klaus Linde et al., "Acupuncture for Patients with Migraine: A Randomized Controlled Trial," *JAMA* 293, no. 17 (May 4, 2005): 2118–25; Dieter Melchart et al., "Acupuncture in Patients with Tension-Type Headache: Randomised Controlled Trial," *BMJ* 331 (2005): 376.

33 *Aldous Huxley once wrote:* Dr. Felix Mann, with foreword by Aldous Huxley, *Acupuncture: Cure of Many Diseases* (London: Pan Books, 1971), v.

33 *but the manifestation of a "superplacebo":* Haake et al., "German Acupuncture Trials."

34 *In a study that pooled together:* Matias Vested Madsen et al., "Acupuncture Treatment for Pain: Systematic Review of Randomised Clinical Trials with Acupuncture, Placebo Acupuncture, and No Acupuncture Groups," *BMJ* 338 (2009): a3115, https://www.bmj.com/content/338/bmj.a3115.

34 *Another, more recent, roundup of studies:* Andrew J. Vickers et al., "Acupuncture for Chronic Pain: Individual Patient Data Meta-Analysis," *Archives of Internal Medicine* 172, no19 (October 22, 2012): 1444–53, http://www.ncbi.nlm.nih.gov/pmc/articles/PMC3658605/.

34 *Studies of chiropractic and osteopathic spinal manipulations show:* S. M. Rubinstein, "Spinal Manipulative Therapy for Chronic Low-Back Pain," *Spine* 36, no. 13 (June 2011): E825–46; John C. Licciardone, "Osteopathic Manipulative Treatment for Low Back Pain: A Systematic Review and Meta-Analysis of Randomized Controlled Trials," *BMC Musculoskeletal Disorders* 6 (2005): 43, https://www.ncbi.nlm.nih.gov/pmc/articles/PMC1208896/.

34 *theories behind the practice:* "How Does Homeopathy Work?," British Homeopathic Association, https://www.britishhomeopathic.org/evidence/how -does-homeopathy-work/.

34 *a 2005 review of 110 studies of homeopathy:* Aijing Shang et al., "Are the Clinical Effects of Homoeopathy Placebo Effects? Comparative Study of Placebo-Controlled Trials of Homoeopathy and Allopathy," *Lancet* 366, no. 9487 (August 27–September 2, 2005): 726–32.

35 *Some preliminary studies show that Reiki and other forms:* Anita Catlin and R. Taylor-Ford, "Investigation of Standard Care Versus Sham Reiki Placebo Versus Actual Reiki Therapy to Enhance Comfort and Well-Being in a Chemotherapy Infusion Center," *Oncology Nursing Forum* 38, no. 3 (May 2011): E212–20; Susan Thrane and Susan M. Cohen, "Effect of Reiki Therapy on Pain and Anxiety in Adults: An In-Depth Literature Review of Randomized Trials with Effect Size Calculations," *Pain Management Nursing* 15, no. 4 (December 2014): 897–908; Shamini Jain et al., "Complementary Medicine for Fatigue and Cortisol Variability in Breast Cancer Survivors: A Randomized Controlled Trial," *Cancer* 118, no. 3 (February 1, 2012): 777–87; Janice Post-White et al.,"Therapeutic Massage and Healing Touch Improve Symptoms in Cancer," *Integrative Cancer Therapies* 2, no. 4 (December 2003): 332–44.

35 *health authorities in Germany decided:* K. Trinczek, "Reimbursement for Acupuncture Treatments in the German Statutory Health Insurance System," *Forschende Komplementämedizin* 22, no. 2 (2015): 118–23.

36 *the UK's National Institute for Health and Care Excellence reversed course:* "NICE Publishes Updated Advice on Treating Low Back Pain," National Institute for Health and Care Excellence, November 30, 2016, https://www.nice.org .uk/news/article/nice-publishes-updated-advice-on-treating-low-back-pain.

36 *still recommends it for chronic tension headaches and migraines:* National Health Service: Acupuncture, https://www.nhs.uk/conditions/acupuncture/.

36 *Chiropractors and other manual therapists similarly struggle to find their sugar pill:* Mark J. Hancock et al., "Selecting an Appropriate Placebo for a Trial of Spinal Manipulative Therapy," *Australian Journal of Physiotherapy* 52, no. 2 (2006):135–38, https://cdn.bodyinmind.org/wp-content/uploads/Aust -J-Physiother-2006-Hancock.pdf.

37 *Dispenza tells the story of getting hit by an SUV:* Joe Dispenza, "How I Healed Myself After Breaking 6 Vertebrae: The Placebo Effect in Action," May 23, 2014, https://www.healyourlife.com/how-i-healed-myself-after-breaking -6-vertebrae.

Chapter Four: The Pharmacy Within

43 *were occasionally cured by nothing more than a "shock to the mind":* Pierre Janet, *Psychological Healing: A Historical and Clinical Study*, vol. 1 (London: George Allen & Unwin, 1925). 49.

43   *Much later, in 1807, Thomas Jefferson wrote:* Thomas Jefferson, *Memoirs, Correspondence and Private Papers of Thomas Jefferson, Late President of the United States* (London: Henry Colburn and Richard Bentley, 1829), 95, https://books.google.com/books?id=gfk5AAAAcAAJ&printsec=frontcover&source=gbs_ge_summary_r&cad=0#v=onepage&q&f=false.

43   *The word* placebo *is said to have originated:* Jeff Aronson, "Please, Please Me," *BMJ* 318, no. 7185 (March 13,1999): 716, https://www.ncbi.nlm.nih.gov/pmc/articles/PMC1115150/.

43   *told of administering a patient a dose of mustard powder:* Catherine Kerr et al., "William Cullen and a Missing Mind-Body Link in the Early History of Placebos," *Journal of the Royal Society of Medicine* 101, no. 2 (February 2008): 89–92, https://www.ncbi.nlm.nih.gov/pmc/articles/PMC2254457/.

44   *There, in makeshift military field hospitals, soldiers arrived:* Henry K. Beecher, "Relationship of Significance of Wound to Pain Experienced," *JAMA* 161, no. 17 (1956): 1609–13; Henry K. Beecher, "Pain in Men Wounded in Battle," *Annals of Surgery* 123 (1946): 96–105, https://www.ncbi.nlm.nih.gov/pmc/articles/PMC1803463/pdf/annsurg01382-0108.pdf.

44   *Publishing his findings in 1955:* Henry K. Beecher, "The Powerful Placebo," *JAMA* 159, no. 17 (1955): 1602–6, http://jgh.ca/uploads/Psychiatry/Links/beecher.pdf.

45   *In 1997, a pair of German medical statisticians concluded:* Gunver S. Kienle and Helmut Kiene, "The Powerful Placebo Effect: Fact or Fiction?" *Journal of Clinical Epidemiology* 50, no. 12 (December 1997): 1311–18, https://pdfs.semanticscholar.org/d56f/d32831c392a90a69cd7a76db8199a9da6607.pdf.

46   *In 1970, the FDA mandated for the first time:* Stephen P. Glasser, *Essentials of Clinical Research* (New York: Springer, 2008), 115.

46   *in the late 1970s, two neuroscientists at the University of California:* Conversations in June 2016 with Howard Fields; Jon D. Levine et al., "The Narcotic Antagonist Naloxone Enhances Clinical Pain," *Nature* 272, no. 5656 (April 27, 1978): 826–27; Jon D. Levine et al., "The Mechanisms of Placebo Analgesia," *Lancet* 2, no. 8091 (September 23, 1978): 654–57.

47   *In follow-up experiments, Fields and Levine determined:* Jon D. Levine et al., "Analgesic Responses to Morphine and Placebo in Individuals with Postoperative Pain," *Pain* 10 (1981): 379–89.

47   *Scientists at the University of British Columbia sent radioactive tracers:* Raul de la Fuente-Fernández et al., "Expectation and Dopamine Release: Mechanism of the Placebo Effect in Parkinson's Disease," *Science* 293, no. 5532 (August 10, 2001): 1164–66.

48   *The brain's endocannabinoids:* Fabrizio Benedetti et al., "Nonopioid Placebo Analgesia Is Mediated by CB1 Cannabinoid Receptors," *Nature Medicine* 17, no. 10 (October 2, 2011): 1228–30.

48   *he told the* New Yorker *in 2011:* Michael Specter, "The Power of Nothing,"

December 12, 2011, https://www.newyorker.com/magazine/2011/12/12
/the-power-of-nothing.

49 *For the 2008 study:* Ted J. Kaptchuk et al., "Components of Placebo Effect:
Randomised Controlled Trial in Patients with Irritable Bowel Syndrome,"
BMJ 336, no. 7651 (May 3, 2008): 999–1003, https://www.ncbi.nlm.nih
.gov/pmc/articles/PMC2364862/.

49 *Earlier nonplacebo studies had shown that:* Zelda Di Blasi et al. "Influence of
Context Effects on Health Outcomes: A Systematic Review," *Lancet* 357,
no. 9258 (March 10, 2001): 757–62; John M. Kelley et al., "The Influence
of the Patient-Clinician Relationship on Healthcare Outcomes: A Systematic Review and Meta-Analysis of Randomized Controlled Trials," *PLoS
One* 9, no. 4 (April 9, 2014): e94207, http://journals.plos.org/plosone
/article?id=10.1371/journal.pone.0094207.

50 *In one of Kaptchuk's earlier articles:* Ted J. Kaptchuk, "The Placebo Effect in
Alternative Medicine: Can the Performance of a Healing Ritual Have Clin-
ical Significance?" *Annals of Internal Medicine* 136 (2002): 817–825, http://
belmont.bme.umich.edu/wp-content/uploads/sites/377/2018/02/1
-The-Placebo-Effect-in-Alternative-Medicine-Can-the-Performance-of-a
-Healing-Ritual-Have-Clinical-Significance.pdf.

50 *Studies have shown that undergoing vertebroplasty for spinal fractures,* etc: Rach-
elle Buchbinder et al., "A Randomized Trial of Vertebroplasty for Painful
Osteoporotic Vertebral Fractures," *New England Journal of Medicine* 361
(2009): 557–68; David F. Kallmes et al., "A Randomized Trial of Vertebroplasty for Osteoporotic Spinal Fractures," *New England Journal of Medicine*
361 (2009): 569–79; J. Bruce Moseley, "A Controlled Trial of Arthroscopic
Surgery for Osteoarthritis of the Knee," *New England Journal of Medicine*
347 (2002): 81–88; Martin B. Leon et al., "A Blinded, Randomized, Placebo-
Controlled Trial of Percutaneous Laser Myocardial Revascularization to
Improve Angina Symptoms in Patients with Severe Coronary Disease,"
Journal of the American College of Cardiology 46, no. 10 (November 15,
2005): 1812–19; Rasha Al-Lamee et al., "Percutaneous Coronary Interven-
tion in Stable Angina (ORBITA): A Double-Blind, Randomised Controlled
Trial," *Lancet* 391, no. 10115 (January 6, 2018): 31–40; David J. Beard et al.,
"Arthroscopic Subacromial Decompression for Subacromial Shoulder Pain
(CSAW): A Multicentre, Pragmatic, Parallel Group, Placebo-Controlled,
Three-Group, Randomised Surgical Trial," *Lancet* 391, 10118 (January 27,
2018): 329–38.

51 *he documented a placebo response within a single neuron of a patient's brain:*
Fabrizio Benedetti et al., "Placebo-Responsive Parkinson Patients Show
Decreased Activity in Single Neurons of Subthalamic Nucleus," *Nature
Neuroscience* 7, no. 6 (June 2004): 587–88.

53 *In 2003, French oncologists rounded up seven cancer studies:* Gisèle Chvetzoff
and Ian F. Tannock, "Placebo Effects in Oncology," *Journal of the National*

Cancer Institute 95, no. 1 (January 1, 2003): 19–29, http://jnci.oxfordjournals
.org/content/95/1/19.full.

55   *acupuncturists enjoy the highest rate of US physician referral:* John A. Astin
et al., "A Review of the Incorporation of Complementary and Alternative
Medicine by Mainstream Physicians," *Archives of Internal Medicine* 158,
no. 21 (November 23, 1998): 2303–10, https://jamanetwork.com/journals
/jamainternalmedicine/fullarticle/210591.

56   *As Ted Kaptchuk observed:* Kaptchuk, "Placebo Effect in Alternative Medicine," 817–25.

56   *Two recent studies have shown:* Cláudia Carvalho et al., "Open-Label Placebo
Treatment in Chronic Low Back Pain: A Randomized Controlled Trial,"
Pain 157, no. 12 (December 2016): 2766–72; Ted J. Kaptchuk et al., "Pla-
cebos Without Deception: A Randomized Controlled Trial in Irritable
Bowel Syndrome," *PLoS One* 5, no. 12 (2010): e15591.

58   *One study found that a placebo pill significantly reduced:* Sabine Vits et al.,
"Cognitive Factors Mediate Placebo Responses in Patients with House
Dust Mite Allergy," *PLoS One* 8, no. 11 (2013): e79576.

58   *Studies have exposed mildly asthmatic people:* L. Y. Liu et al., "School Examinations Enhance Airway Inflammation to Antigen Challenge," *American Journal of Respiratory and Critical Care Medicine* 165, no. 8 (April 15,
2002): 1062–67; Edith Chen and Gregory E. Miller, "Stress and Inflammation in Exacerbations of Asthma," *Brain, Behavior, and Immunity* 21, no. 8
(November 2007); 993–99, https://www.ncbi.nlm.nih.gov/pmc/articles
/PMC2077080/.

59   *The handful of studies that have been done on acupuncture and allergic rhinitis:*
Benno Brinkhaus et al., "Acupuncture in Patients with Seasonal Allergic
Rhinitis: A Randomized Trial," *Annals of Internal Medicine* 158, no. 4 (February 19, 2013): 225–34; Miriam Ortiz et al., "Autonomic Function in Seasonal Allergic Rhinitis and Acupuncture—an Experimental Pilot Study
within a Randomized Trial," *Forschende Komplementämedizin* 22, no. 2
(2015): 85–92, https://www.karger.com/Article/FullText/381086.

59   *The American Academy of Otolaryngology:* "AAO–HNSF Releases Guideline
on Allergic Rhinitis," *American Family Physician* 92, no. 10 (November 15,
2015): 942–44, https://www.aafp.org/afp/2015/1115/p942.html.

Chapter Five: Healing Partners

62   *some 84 percent of us do:* Janet K. Freburger et al., "The Rising Prevalence
of Chronic Low Back Pain," *Archives of Internal Medicine* 169, no. 3 (2009):
251–58, https://www.ncbi.nlm.nih.gov/pmc/articles/PMC4339077/.

64   *Both spinal fusion and decompression surgery:* Peter Whoriskey and Dan Keating, "Spinal Fusions Serve as Case Study for Debate over When Certain
Surgeries Are Necessary," *Washington Post*, October 27, 2013; John Car-

reyrou and Tom McGinty, "Top Spine Surgeons Reap Royalties, Medicare Bounty," *Wall Street Journal,* December 20, 2010.

69 *Mindfulness expert Jon Kabat-Zinn once observed:* Bill Moyers, "For Stress Reduction, Meditate! An Expert Explains Why Meditation Can Help Reduce Stress," *Psychology Today,* July 1, 1993.

78 *Richard Gracely at the University of North Carolina calls:* Richard H. Gracely, "Charisma and the Art of Healing: Can Nonspecific Factors Be Enough?," *Proceedings of the Ninth World Congress on Pain,* vol. 16, ed. Zsuzsanna Wiesenfeld-Hallin et al. (IASP Press, 2000); conversation with Richard Gracely, August 2016.

Chapter Six: My Back Is Out

80 *The first chiropractic treatment took place in the town of Davenport:* Joseph C. Keating Jr. et al., *Chiropractic History: A Primer* (Rock Island, IL: Association for the History of Chiropractic, 2004), https://www.brianesty.com /bodywork/PDF/Chiropractic%20History.pdf.

80 *The whole idea was famously debunked:* Benjamin Franklin, *Animal Magnetism: Report of Dr. Franklin and Other Commissioners, Charged by the King of France with the Examination of the Animal Magnetism as Practised at Paris* (Philadelphia: Perkins, 1837).

81 *According to Palmer, Lillard told him:* Daniel David Palmer, *The Science, Art and Philosophy of Chiropractic* (Portland, OR: Portland Printing House, 1910).

81 *Palmer described Lillard as "deaf":* Daniel David Palmer, *The Chiropractor's Adjuster* (Portland, OR: Portland Printing House, 1910).

81 *he wrote in 1910:* Ibid.

81 *including one with unspecified "heart trouble":* Palmer, *Science, Art and Philosophy of Chiropractic.*

82 *In the "done by hand" therapy of* chiropractic: Myron D. Brown, "Old Dad Chiro: His Thoughts, Words, and Deeds," *Journal of Chiropractic Humanities* 16, no. 1 (December 2009): 57–75, https://www.ncbi.nlm.nih.gov/pmc /articles/PMC3342800/.

82 *He claimed that "95 percent of all diseases":* Palmer, *Chiropractor's Adjuster.*

82 *by 1902, fifteen chiropractors had graduated from Palmer College:* New York State Chiropractors Association, "Chiropractic History," https://www .nysca.com/patients/chiropractic_history.asp.

82 *Palmer wasn't the only starry-eyed medical entrepreneur:* Andrew Taylor Still, *Osteopathy, Research and Practice* (St. Paul, MN: Press of the Pioneer Company, 1910); Carol Trowbridge, *Andrew Taylor Still, 1828–1917* (Kirksville, MO: Thomas Jefferson University Press, 1991).

83 *The practice appears throughout history:* Erland Pettman, "A History of Manipulative Therapy," *Journal of Manual & Manipulative Therapy* 15,

no. 3 (2007): 165–74, https://www.ncbi.nlm.nih.gov/pmc/articles/PMC2565620/.

84 *hundreds of them were sued for practicing medicine without a license:* Ted J. Kaptchuk et al., "Chiropractic Origins, Controversies, and Contributions," *Archives of Internal Medicine* 158 (1998): 2215–24; National Institute of Chiropractic Research, DD Palmer's Lifeline, http://www.chiro.org /Plus/History/Persons/PalmerDD/DDs_LIFELINE_Chronology.PDF.

84 *the powerful American Medical Association tried to put a stop:* "Chiropractic Condemned," *JAMA* 208, no. 2 (1969): 352.

84 *In 1987, a federal judge ruled in their favor:* "U.S. Judge Finds Medical Group Conspired Against Chiropractors," *New York Times,* August 29, 1987.

84 *the profession continued to grow:* American Chiropractic Association, "Key Facts About the Chiropractic Profession," https://www.acatoday.org /Patients/Why-Choose-Chiropractic/Key-Facts; R. A. Cooper and S. J. Stoflet, "Trends in the Education and Practice of Alternative Medicine," *Health Affairs* 15, no. 3 (1996): 226–38.

87 *neck manipulations carry a small risk of complications:* Michael John Haynes et al., "Assessing the Risk of Stroke from Neck Manipulation: A Systematic Review," *International Journal of Clinical Practice* 66, no. 10 (October 2012): 940–47, https://www.ncbi.nlm.nih.gov/pmc/articles/PMC3506737/.

87 *Studies have shown, however, that spinal manipulations of the sort Dr. Bell did:* E. L. Hurwitz et al., "Manipulation and Mobilization of the Cervical Spine: A Systematic Review of the Literature," *Spine* 21, no. 15 (August 1, 1996): 1746–59.

87 *a compilation of all the higher-quality research:* S. M. Rubinstein, "Spinal Manipulative Therapy for Chronic Low-Back Pain," *Spine* 36, no. 13 (June 2011): E825–46.

87 *Both the American College of Physicians and the American Pain Society recommend:* Roger Chou, "Nonpharmacologic Therapies for Low Back Pain: A Systematic Review for an American College of Physicians Clinical Practice Guideline," *Annals of Internal Medicine* 166, no. 7 (April 4, 2017): 493–505, http://annals.org/aim/fullarticle/2603230/nonpharmacologic-therapies -low-back-pain-systematic-review-american-college-physicians; or https:// www.acponline.org/acp-newsroom/american-college-of-physicians-issues -guideline-for-treating-nonradicular-low-back-pain.

89 *In the few studies that have compared one type of spinal manipulation or massage to another:* Maria A. Hondras et al., "A Randomized Controlled Trial Comparing 2 Types of Spinal Manipulation and Minimal Conservative Medical Care for Adults 55 Years and Older with Subacute or Chronic Low Back Pain," *Journal of Manipulative and Physiological Therapeutics* 32, no. 5 (June 2009): 330–43; Chad Cook et al., "Early Use of Thrust Manipulation Versus Non-Thrust Manipulation: A Randomized Clinical Trial," *Manual Therapy* 18, no. 3 (June 2013): 191–98; Daniel C. Cherkin et al., "A Com-

parison of the Effects of 2 Types of Massage and Usual Care on Chronic Low Back Pain: A Randomized, Controlled Trial," *Annals of Internal Medicine* 155, no. 1 (July 5, 2011): 1–9, https://www.ncbi.nlm.nih.gov/pmc /articles/PMC3570565/.

89 *when a practitioner was allowed to evaluate a patient:* Peter Kent et al., "Does Clinician Treatment Choice Improve the Outcomes of Manual Therapy for Non-Specific Low Back Pain: A Meta-Analysis," *Journal of Manipulative and Physiological Therapeutics* 28, no. 5 (June 2005): 312–22; Adit Chiradejnant et al., "Efficacy of 'Therapist-Selected' Versus 'Randomly Selected' Mobilisation Techniques for the Treatment of Low Back Pain: A Randomised Controlled Trial," *Australian Journal of Physiotherapy* 49, no. 4 (2003): 233–41.

90 *manual therapists are not good at using their hands:* Michael A. Seffinger et al., "Reliability of Spinal Palpation for Diagnosis of Back and Neck Pain: A Systematic Review of the Literature," *Spine* 29, no. 19 (October 1, 2004): E413–25; Peter A. Huijbregts, "Spinal Motion Palpation: A Review of Reliability Studies," *Journal of Manual & Manipulative Therapy* 10, no. 1 (2002): 24–39; Daniel L. Riddle and Janet K. Freburger, "Evaluation of the Presence of Sacroiliac Joint Region Dysfunction Using a Combination of Tests: A Multicenter Inter-tester Reliability Study," *Physical Therapy* 82, no. 8 (August 2002): 772–81.

90 *In one study done at the Arthritis Research Center in Wichita:* Frederick Wolfe et al., "The Fibromyalgia and Myofascial Pain Syndromes: A Preliminary Study of Tender Points and Trigger Points in Persons with Fibromyalgia, Myofascial Pain Syndrome and No Disease," *Journal of Rheumatology* 19, no. 6 (June 1992): 944–51.

90 *researchers at the University of Alberta used MRI machines:* Gregory N. Kawchuk et al., "Real-Time Visualization of Joint Cavitation," *PLoS One* 10, no. 4 (April 15, 2015): e0119470, http://journals.plos.org/plosone /article?id=10.1371/journal.pone.0119470.

91 *Bialosky sometimes refers to this common underlying mechanism as a placebo effect:* Joel E. Bialosky et al., "Placebo Response to Manual Therapy: Something Out of Nothing?" *Journal of Manual & Manipulative Therapy* 19, no. 1 (February 2011): 11–19, https://www.ncbi.nlm.nih.gov/pmc/articles /PMC3172952/.

91 *our brain's response to the physical sensations:* https://www.ncbi.nlm.nih.gov /pmc/articles/PMC2775050/.; Joel E. Bialosky et al., "The Mechanisms of Manual Therapy in the Treatment of Musculoskeletal Pain: A Comprehensive Model," *Manual Therapy* 14, no. 5 (October 2009): 531–38, https:// www.ncbi.nlm.nih.gov/pmc/articles/PMC4976880/; Richard Edward Nyberg and A. Russell Smith Jr., "The Science of Spinal Motion Palpation: A Review and Update with Implications for Assessment and Intervention," *Journal Manual & Manipulative Therapy* 21, no. 3 (August 2013): 160–67, https://www.ncbi.nlm.nih.gov/pmc/articles/PMC3744849/.

92   *semi-organized group of acupuncture-curious scientists:* Helene M. Langevin et al., "Paradoxes in Acupuncture Research: Strategies for Moving Forward," *Evidence-Based Complementary and Alternative Medicine* 2011, Article ID 180805, https://www.hindawi.com/journals/ecam/2011/180805/.

Chapter Seven: This Feeling in My Body

93   *"three pounds of the most complex material":* David Eagleman, *Incognito: The Secret Lives of the Brain* (New York: Vintage Books, 2011), https://www.eagleman.com/incognito/excerpts.

94   *Thanassis Martinos donated $20 million:* "$20 Million Gift Will Advance Imaging Technology at New MIT-Harvard Center," *MIT News,* May 19, 1999, http://news.mit.edu/1999/martinos.

94   *Martinos Center scientists have gotten some $25 million:* Conversation with Vitaly Napadow, March 2018.

94   *In blog posts in 2013 and 2015, Steven Novella:* Steven Novella, "Acupuncture Doesn't Work," *Science-Based Medicine,* June 19, 2013, https://sciencebasedmedicine.org/acupuncture-doesnt-work/.

96   *A study done at the University of California, Irvine:* Z. H. Cho et al., "New Findings of the Correlation Between Acupoints and Corresponding Brain Cortices Using Functional MRI," *Proceedings of the National Academy of Sciences U.S.A.* 95, no. 5 (March 3, 1998): 2670–73, https://www.ncbi.nlm.nih.gov/pmc/articles/PMC19456/.

97   *Merely closing one's eyes can, paradoxically, activate:* T. Brandt, "How to See What You Are Looking for in fMRI and PET—or the Crucial Baseline Condition," *Journal of Neurology* 253, no. 5 (May 2006): 551–55.

97   *In 2006, five of the eight researchers retracted the paper:* Z. H. Cho et al., "Retraction. New Findings of the Correlation Between Acupoints and Corresponding Brain Cortices Using Functional MRI," *Proceedings of the National Academy of Sciences U.S.A.* 103, no. 27 (July 5, 2006): 10527.

97   *studies that he and other acupuncture researchers have done with these machines:* Wenjing Huang et al., "Characterizing Acupuncture Stimuli Using Brain Imaging with fMRI—a Systematic Review and Meta-Analysis of the Literature," *PLoS One* 7, no. 4 (2012): e32960, https://www.ncbi.nlm.nih.gov/pmc/articles/PMC3322129/.

99   *he was part of a study that used radioactive PET scans:* Richard E. Harris et al., "Traditional Chinese Acupuncture and Placebo (Sham) Acupuncture Are Differentiated by Their Effects on Mu-Opioid Receptors (MORs)," *Neuroimage* 47, no. 3 (September 2009): 1077–85, https://www.ncbi.nlm.nih.gov/pmc/articles/PMC2757074/.

100 *When one type of exercise is compared against another:* Bruno Tirotti Saragiotto et al., "Motor Control Exercise for Chronic Non-Specific Low-Back Pain," *Cochrane Database of Systematic Reviews* 1 (January 8, 2016): CD012004;

Edward Roddy et al., "Aerobic Walking or Strengthening Exercise for Osteoarthritis of the Knee? A Systematic Review," *Annals of the Rheumatic Diseases* 64, no. 4 (April 2005): 544–48.

100 *It is assumed that, for example, those with spinal instability:* Conversation with Luciana Macedo, May 2018.

Chapter Eight: Brain Pain

104 *50 million of us, or one in five adults—suffer:* Richard L. Nahin, "Estimates of Pain Prevalence and Severity in Adults: United States, 2012," *Journal of Pain* 16, no. 8 (August 2015): 769–80, https://www.ncbi.nlm.nih.gov /pmc/articles/PMC4562413/.

105 *According to studies, CBT helps roughly 40 percent:* For a review of CBT efficacy for chronic pain, see Dawn M. Ehde et al., "Cognitive-Behavioral Therapy for Individuals with Chronic Pain Efficacy: Innovations, and Directions for Research," *American Psychologist*, February–March 2014, https://www.apa.org/pubs/journals/releases/amp-a0035747.pdf.

105 *Mindfulness-based stress reduction:* Daniel C. Cherkin et al., "Effect of Mindfulness-Based Stress Reduction vs. Cognitive Behavioral Therapy or Usual Care on Back Pain and Functional Limitations in Adults with Chronic Low Back Pain: A Randomized Clinical Trial," *JAMA* 315, no. 12 (2016): 1240–49; Lance M. McCracken and Kevin E. Vowles, "Acceptance and Commitment Therapy and Mindfulness for Chronic Pain: Model, Process, and Progress," *American Psychologist* 69, no. 2 (2014): 178–87.

105 *An even newer approach, called emotional awareness and expression therapy:* Mark A. Lumley et al., "Emotional Awareness and Expression Therapy, Cognitive Behavioral Therapy, and Education for Fibromyalgia: A Cluster-Randomized Controlled Trial," *Pain* 158, no. 12 (December 2017): 2354–63, https://static1.squarespace.com/static/583c6eea6b8f5b152d3bdb4b/t /59b8559fedaed8ae88e62a03/1505252768600/0.pdf.

106 *tangled bit of locution in the respected* Nature Reviews Neuroscience*:* Erin D. Milligan and Linda R. Watkins, "Pathological and Protective Roles of Glia in Chronic Pain," *Nature Reviews Neuroscience* 10, no. 1 (January 2009): 23–36.

108 *In an early nineties study . . . a team of psychologists at Baylor:* Timothy L. Bayer et al., "Situational and Psychophysiological Factors in Psychologically Induced Pain," *Pain* 44, no. 1 (January 1991): 45–50.

108 *a bizarre real-world example of this kind of sourceless pain:* J. P. Fisher et al., "Minerva," *BMJ* 310 (1995): 70.

109 *doctors at Hoag Memorial Hospital in California did a study that stunned nearly every spine doctor:* Maureen C. Jensen et al., "Magnetic Resonance Imaging of the Lumbar Spine in People Without Back Pain," *New England Journal of Medicine* 331, no. 2 (July 14, 1994): 69–73.

109 *revealed that even pain-free twenty-year-olds:* Waleed Brinjikji et al., "Systematic Literature Review of Imaging Features of Spinal Degeneration in Asymptomatic Populations," *American Journal of Neuroradiology* 36, no. 4 (April 2015): 811–16; https://www.ncbi.nlm.nih.gov/pmc/articles /PMC4464797/.

109 *Between 20 and 60 percent of people with no knee problems:* Paul Creamer and Marc C. Hochberg, "Why Does Osteoarthritis of the Knee Hurt—Sometimes?," *British Journal of Rheumatology* 36 (1997): 721–28.

109 *a recent study found that about half of pain-free jaws:* Jens C. Türp et al., "Disc Displacement, Eccentric Condylar Position, Osteoarthritis—Misnomers for Variations of Normality? Results and Interpretations from an MRI Study in Two Age Cohorts," *BMC Oral Health* 16 (2016):124, https:// www.ncbi.nlm.nih.gov/pmc/articles/PMC5114831/.

109 *Conversely, between 10 and 15 percent of people with relatively normal knee X-rays:* Creamer and Hochberg, "Why Does Osteoarthritis of the Knee Hurt?," 721–28.

110 *Back pain experts now consider about 90 percent of all cases to be "nonspecific":* Chris Maher et al., "Non-specific Low Back Pain," *Lancet* 389, no. 10070 (February 18, 2017): 736–47, https://www.thelancet.com/journals/lancet /article/PIIS0140-6736(16)30970-9/fulltext.

110 *why spinal surgery is often ineffective for back problems:* Zafeer Baber and Michael A. Erdek, "Failed Back Surgery Syndrome: Current Perspectives," *Journal of Pain Research* 9 (2016): 979–87.

110 *why vertebroplasty . . . has been shown to be no more effective:* Cristina E. Firanescu et al., "Vertebroplasty Versus Sham Procedure for Painful Acute Osteoporotic Vertebral Compression Fractures (VERTOS IV): Randomised Sham Controlled Clinical Trial," *BMJ* 361 (2018): k1551, https:// www.bmj.com/content/361/bmj.k1551.

110 *A driving feature of much chronic pain:* For more information on this, see "What Is Central Sensitization?," Institute for Chronic Pain, http:// www.instituteforchronicpain.org/understanding-chronic-pain/what-is -chronic-pain/central-sensitization; and Paul Ingraham, "Central Sensitization in Chronic Pain," PainScience.com, February 19, 2017, https:// www.painscience.com/articles/central-sensitization.php.

111 *his group in Adelaide recently tested people with self-reported "back stiffness":* Tasha R. Stanton et al., "Feeling Stiffness in the Back: A Protective Perceptual Inference in Chronic Back Pain," *Scientific Reports* 7 (2017): 9681, https://www.nature.com/articles/s41598-017-09429-1.

111 *Similar tendencies have been observed in patients with irritable bowel syndrome and fibromyalgia:* G. Nicholas Verne and Donald D. Price, "Irritable Bowel Syndrome as a Common Precipitant of Central Sensitization," *Current Rheumatology Reports* 4, no. 4 (August 2002): 322–28; Thorsten Giesecke et al., "Evidence of Augmented Central Pain Processing in Idiopathic

Chronic Low Back Pain," *Arthritis & Rheumatism* 50, no. 2 (February 2004): 613–23.

112 *the brains of people whose low-back pain becomes chronic:* Marwan N. Baliki et al., "Corticostriatal Functional Connectivity Predicts Transition to Chronic Back Pain," *Nature Neuroscience* 15, no. 8 (August 2012): 1117–19; Javeria A. Hashmi et al., "Shape Shifting Pain: Chronification of Back Pain Shifts Brain Representation from Nociceptive to Emotional Circuits," *Brain* 136, pt. 9 (September 2013): 2751–68; Etienne Vachon-Presseau et al., "Corticolimbic Anatomical Characteristics Predetermine Risk for Chronic Pain," *Brain* 139, no. 7 (July 2016): 1958–70.

113 *predisposition to react to pain in unhelpful ways due to genes:* For more on what scientists do know, see Katerina Zorina-Lichtenwalter et al., "Genetic Predictors of Human Chronic Pain Conditions," *Neuroscience* 338 (December 3, 2016): 36–62, https://www.sciencedirect.com/science/article/pii/S0306452216301269.

113 *observational studies that show one of the biggest predictors:* Robert R. Edwards et al., "The Role of Psychosocial Processes in the Development and Maintenance of Chronic Pain," *Journal of Pain* 17, no. 9 (suppl.) (September 2016): T70–92, https://www.ncbi.nlm.nih.gov/pmc/articles/PMC5012303/.

Chapter Nine: The Illness of Disease

119 *some eight hundred, with another thirty-five hundred people:* Conversation with Dr. Fred Kaplan, March 2018.

120 *the French physician Guy Patin . . . wrote to a colleague:* Thomas Maeder, "A Few Hundred People Turned to Bone," *Atlantic*, February 1998.

127 *he and his team have isolated the gene responsible for FOP:* Eileen M. Shore et al., "A Recurrent Mutation in the BMP Type I Receptor ACVR1 Causes Inherited and Sporadic Fibrodysplasia Ossificans Progressiva," *Nature Genetics* 38, no. 5 (May 2006): 525–27; Frederick S. Kaplan et al., "Fibrodysplasia Ossificans Progressiva: Mechanisms and Models of Skeletal Metamorphosis," *Disease Models & Mechanisms* 5, no. 6 (November 2012): 756–62.

128 *Those who teach lymphedema therapy insist it's not easy:* Conversation with Steve Norton, president of the Norton School of Lymphatic Therapy, November 2016.

128 *hail from an early twentieth-century osteopath:* William H. Devine, "Chapman's Reflexes and Modern Clinical Applications," http://files.academyofosteopathy.org/convo/2016/Handouts/Devine_SAAOLecture.pdf.

129 *They enrolled forty-six people with this common disorder of the lungs:* Michael E. Wechsler et al., "Active Albuterol or Placebo, Sham Acupuncture, or No Intervention in Asthma," *New England Journal of Medicine* 365 (2011): 119–26, http://www.nejm.org/doi/full/10.1056/NEJMoa1103319#t=article.

130 *skeptics pounced:* David Gorski, "Spin City: Using Placebos to Evaluate Objective and Subjective Responses in Asthma," *Science-Based Medicine,* July 18, 2011, https://sciencebasedmedicine.org/spin-city-placebos-and -asthma/#more-14434; Steven Novella, "Placebo Medicine," August 8, 2012, https://www.youtube.com/watch?v=YXeX7dQNXvY.

131 *Arthur Kleinman . . . drew a distinction between the treatment of* illness *and* disease: Arthur Kleinman et al., "Culture, Illness, and Care: Clinical Lessons from Anthropologic and Cross-Cultural Research," *Annals of Internal Medicine* 88, no. 2 (February 1978): 251–58, https://www.researchgate .net/publication/22514763_Culture_Illness_and_Care_Clinical_Lessons _From_Anthropologic_and_Cross-Cultural_Research; Arthur Kleinman, *The Illness Narratives* (New York: Basic Books, 1988), 4.

132 *list of eight questions:* https://www.med.upenn.edu/gec/user_docs/PDF /Health%20Equity%20and%20Literacy/Kleinman_s_8_Questions.pdf.

132 *anywhere between 5 to 15 percent of people with multiple sclerosis:* Sara Carletto et al., "Treating Post-Traumatic Stress Disorder in Patients with Multiple Sclerosis: A Randomized Controlled Trial Comparing the Efficacy of Eye Movement Desensitization and Reprocessing and Relaxation Therapy," *Frontiers in Psychology* 7 (2016): 526, https://www.ncbi.nlm.nih.gov/pmc /articles/PMC4838623/.

132 *a neurologist told me:* Conversation with Dr. Allen C. Bowling, April 2015.

132 *The incidence of major depression in cancer patients:* Mary Jane Massie, "Prevalence of Depression in Patients with Cancer," *Journal of the National Cancer Institute Monographs* 32 (2004): 57–71.

133 *he had tried and failed to get a person's expectations:* Fabrizio Benedetti et al., "Conscious Expectation and Unconscious Conditioning in Analgesic, Motor, and Hormonal Placebo/Nocebo Responses," *Journal of Neuroscience* 23, no. 10 (May 15, 2003): 4315–23, http://www.jneurosci.org /content/23/10/4315.

133 *Manfred Schedlowski . . . tried something similar with the immune system:* Marion U. Goebel et al., "Behavioral Conditioning of Immunosuppression Is Possible in Humans," *FASEB Journal* 16, no 14 (December 2002): 1869–73.

134 *Yet many conditions are known to be responsive to our positive expectations:* Fabrizio Benedetti, *Placebo Effects: Understanding the Mechanisms in Health and Disease* (Oxford: Oxford University Press, 2014).

134 *fatigue is another of the brain's protective perceptions:* For more information on the science of fatigue, see Alex Hutchinson, "What Is Fatigue?," *New Yorker,* December 12, 2014.

136 *researchers showed that you could do similar things with the secretion of insulin:* Ursula Stockhorst et al., "Classically Conditioned Responses Following Repeated Insulin and Glucose Administration in Humans," *Behavioural Brain Research* 110, no. 1–2 (June 1, 2000): 143–59.

136 *Benedetti found that fake oxygen could tamp down:* Fabrizio Benedetti et al.,

"Critical Life Functions: Can Placebo Replace Oxygen?," *International Review of Neurobiology* 138 (2018): 201–18.

Chapter Ten: The Zen Response

142 *Then in 1980, scientists working with mice:* John M. Williams et al., "Sympathetic Innervation of Murine Thymus and Spleen: Evidence for a Functional Link Between the Nervous and Immune System," *Brain Research Bulletin* 6 (1980): 83–94.

142 *University of Texas researchers revealed that human immune cells:* J. Edwin Blalock and Eric M. Smith, "Human Leukocyte Interferon: Structural and Biological Relatedness to Adrenocorticotropic Hormone and Endorphins," *Proceedings of the National Academy of Sciences U.S.A.* 77 (1980): 4597–5972; J. Edwin Blalock, "The Immune System as a Sensory Organ," *Journal of Immunology* 132 (1984): 1067–70.

142 *spawned the new field of psychoneuroimmunology:* Robert Ader, "On the Development of Psychoneuroimmunology," *European Journal of Pharmacology* 405 (2000): 167–76.

142 *It's now thought that stress . . . affects immune responses in two opposing ways:* Suzanne C. Segerstrom and Gregory E. Miller, "Psychological Stress and the Human Immune System: A Meta-Analytic Study of 30 Years of Inquiry," *Psychological Bulletin* 130, no. 4 (July 2004): 601–30.

144 *tai chi and immune suppression in older adults:* Michael R. Irwin et al., "Effects of a Behavioral Intervention, Tai Chi Chih, on Varicella-Zoster Virus Specific Immunity and Health Functioning in Older Adults," *Psychosomatic Medicine* 65, no. 5 (September–October 2003): 824–30; Michael R. Irwin et al., "Augmenting Immune Responses to Varicella Zoster Virus in Older Adults: A Randomized Controlled Trial of Tai Chi," *Journal of the American Geriatrics Society* 55, no. 4 (April 2007): 511–17.

145 *Irwin showed that older adults with moderate sleep problems:* Michael R. Irwin et al., "Improving Sleep Quality in Older Adults with Moderate Sleep Complaints: A Randomized Controlled Trial of Tai Chi Chih," *Sleep* 31, no. 7 (July 2008): 1001–8.

145 *A more recent study . . . showed that tai chi for sleep problems in breast cancer survivors:* Michael R. Irwin et al., "Tai Chi Chih Compared with Cognitive Behavioral Therapy for the Treatment of Insomnia in Survivors of Breast Cancer: A Randomized, Partially Blinded, Noninferiority Trial," *Journal of Clinical Oncology* 35, no. 23 (August 10, 2017): 2656–65.

145 *other studies:* For a summary of findings on how mind-body therapies affect immune markers, see Julienne E. Bower and Michael R. Irwin, "Mind-Body Therapies and Control of Inflammatory Biology: A Descriptive Review," *Brain, Behavior, and Immunity* 51 (January 2016): 1–11, https://www.ncbi.nlm.nih.gov/pmc/articles/PMC4679419/.

146 *Studies done elsewhere have shown that tai chi can have positive effects:* "Tai Chi and Qi Gong: In Depth," NCCIH, https://nccih.nih.gov/health/taichi /introduction.htm#use.

146 *studies showing that mindfulness meditation classes . . . can lead to similar reductions:* Julienne Bower et al., "Mindfulness Meditation for Younger Breast Cancer Survivors: A Randomized Controlled Trial," *Cancer* 121 (2015): 1231–40; David S. Black et al., "Mindfulness Meditation and Improvement in Sleep Quality and Daytime Impairment Among Older Adults with Sleep Disturbances: A Randomized Clinical Trial," *JAMA Internal Medicine* 175, no. 4 (April 1, 2015): 494–501; J. David Creswell et al., "Mindfulness-Based Stress Reduction Training Reduces Loneliness and Pro-Inflammatory Gene Expression in Older Adults: A Small Randomized Controlled Trial," *Brain, Behavior, and Immunity* 26, no 7 (October 2012): 1095–1101; Julienne Bower et al., "Inflammation and Behavioral Symptoms after Breast Cancer Treatment: Do Fatigue, Depression, and Sleep Disturbance Share a Common Underlying Mechanism?," *Journal of Clinical Oncology* 29, no. 26 (2011): 3517–22.

148 *many people have expressed skepticism:* Sheldon D. Solomon, "Laughing All the Way to the Bank," *Archives of Internal Medicine* 151, no. 2 (1991): 404.

149 *a study done at Northwestern University's Feinberg School of Medicine:* David C. Mohr et al., "A Randomized Trial of Stress Management for the Prevention of New Brain Lesions in MS," *Neurology* 79, no. 5 (July 31, 2012): 412–19, https://www.ncbi.nlm.nih.gov/pmc/articles/PMC3405245/.

151 *a study done in Israel that gave thirty-eight patients with active breast cancer:* Lee Shaashua et al., "Perioperative COX-2 and β-Adrenergic Blockade Improves Metastatic Biomarkers in Breast Cancer Patients in a Phase-II Randomized Trial," *Clinical Cancer Research* 23, no. 16 (August 15, 2017): 4651–61; Rita Haldar et al., "Perioperative Inhibition of β-Adrenergic and COX-2 Signaling in a Clinical Trial in Breast Cancer Patients Improves Tumor Ki-67 Expression, Serum Cytokine Levels, and Pbmcs Transcriptome," *Brain, Behavior, and Immunity* (May 22, 2018), pii: S0889-1591(18)30187-9.

151 *also got similar results in patients with colon cancer:* Conversation with Julie Bower regarding unpublished results, November 2017.

Chapter Eleven: The Emotional Rescue

153 *Bernadette Soubirous:* Ruth Harris, *Lourdes: Body and Spirit in the Secular Age* (London: Penguin Books, 1999); Francis Trochu, *Saint Bernadette Soubirous: 1844–1879* (New York: Pantheon Books, 1958); Oliver Todd, *The Lourdes Pilgrim* (Chelmsford, UK: Matthew James Publishing, 2003).

155 *a pregnant woman named Catherine Latapie:* "List of Approved Lourdes Miracles," MiracleHunter.com, http://www.miraclehunter.com/marian _apparitions/approved_apparitions/lourdes/miracles1.html#latapie.

155 *6 million people a year now come to visit:* Lourdes Tourist Office, https://www.lourdes-infotourisme.com/web/EN/394-history.php.

156 *in 1883, a medical bureau was established:* Steve Fishman, "A Lovely Day for a Miracle," in *Travelers' Tales: France*, ed. James O'Reilly et al. (Berkeley: Publishers Group West, 1995).

156 *another layer was added, an international committee:* "The International Medical Committee of Lourdes," Lourdes Sanctuaire, https://www.lourdes-france.org/en/international-medical-committee-lourdes.

156 *roughly seven thousand people claiming a healing at Lourdes:* Conversation with Dr. Alessandro de Franciscis, September 2015.

158 *Scientists in the emerging field of neurotheology:* David Biello, "Searching for God in the Brain," *Scientific American Mind*, October/November 2007; John Langone, "In Search of the 'God Gene,'" *New York Times*, November 2, 2004.

159 *the most recent miracle . . . an Italian woman named Danila Castelli:* "Danila Castelli: 69th Cure of Lourdes Recognized as Miraculous by a Bishop," Lourdes Sanctuaire, https://www.lourdes-france.org/en/tv-lourdes/conferences/danila-castelli-69th-cure-lourdes-recognized-miraculous-bishop.

159 *She would fall victim to heart palpitations:* "Report Concerning the Case of the Cure of Mrs. Danila Castelli, Born in Pavia, 19/01/1946."

169 *12 million American adults who will suffer from panic attacks:* "Panic Disorder," National Institutes of Mental Health, https://www.nimh.nih.gov/health/statistics/panic-disorder.shtml#part_155948.

169 *the prevailing theory is that they are a reaction:* Alicia E. Meuret et al., "Do Unexpected Panic Attacks Occur Spontaneously?," *Biological Psychiatry* 70, no. 10 (November 15, 2011): 985–91; "Understanding the Brain's 'Suffocation Alarm,'" *Science Daily*, December 1, 2014, https://www.sciencedaily.com/releases/2014/12/141201090430.htm.

170 *Popularized in the eighties by a Canadian endocrinologist:* Otto Kuchel, "Pseudopheochromocytoma," *Hypertension* 7, no. 1 (January–February 1985): 151–58.

170 *Dr. Samuel Mann . . . believes pseudo-pheochromocytoma is driven:* Samuel J. Mann, "Severe Paroxysmal Hypertension (Pseudopheochromocytoma): Understanding the Cause and Treatment," *Archives of Internal Medicine* 159, no. 7 (April 12, 1999): 670–74; Samuel J. Mann, "Severe Paroxysmal Hypertension (Pseudopheochromocytoma)," *Current Hypertension Reports* 10, no. 1 (February 2008): 12–18.

171 *as an estimated 13 to 34 percent of Italian women have:* Beatrice Castelli et al., "Prevalence of Child Sexual Abuse: A Comparison Among 4 Italian Epidemiological Studies," *Pediatria Medica e Chirurgica* 37, no. 2 (September 28, 2015): pmc.2015.114; UN Global Database on Violence Against Woman, http://evaw-global-database.unwomen.org/en/countries/europe/italy.

Chapter Twelve: All in My Head

174 *Heinroth wrote:* Howard Steinberg et al., "Johann Christian August Heinroth: Psychosomatic Medicine Eighty Years Before Freud," *Psychiatria Danubina* 25, no. 1 (March 2013): 11–16.

177 *Between 1 and 9 percent of people showing up at neurology offices:* Jon Stone et al., "Systematic Review of Misdiagnosis of Conversion Symptoms and 'Hysteria,'" *BMJ* 331 (2005): 989, https://www.bmj.com/content /331/7523/989.

185 *perhaps as many as 50 percent of them had damage:* Anne Louise Oaklander et al., "Objective Evidence That Small-Fiber Polyneuropathy Underlies Some Illnesses Currently Labeled as Fibromyalgia," *Pain* 154, no. 11 (November 2013): 2310–16, https//doi.org.10.1016/j.pain.2013.06.001; Daniel J. Clauw, "What Is the Meaning of 'Small Fiber Neuropathy' in Fibromyalgia?," *Pain* 156 (2015): 2115–16.

185 *researchers are now investigating how . . . disordered immune cells and patterns:* Jose G. Montoya et al., "Cytokine Signature Associated with Disease Severity in Chronic Fatigue Syndrome Patients," *Proceedings of the National Academy of Sciences U.S.A.* 114, no. 34 (August 22, 2017): E7150–58; Dorottya Nagy-Szakal et al., "Fecal Metagenomic Profiles in Subgroups of Patients with Myalgic Encephalomyelitis/Chronic Fatigue Syndrome," *Microbiome* 5 (2017): 44.

185 *Arthur Shapiro . . . successfully advanced the idea that Tourette's:* Giovanna Breu, "Dr. Arthur Shapiro's Battle Against Tourette Syndrome Gets a Boost from TV's Quincy," *People*, March 30, 1981.

Chapter Thirteen: Something to Believe In

198 *2007 paper written by a panel of twenty scientists:* James W. Fawcett et al., "Guidelines for the Conduct of Clinical Trials for Spinal Cord Injury as Developed by the ICCP Panel: Spontaneous Recovery after Spinal Cord Injury and Statistical Power Needed for Therapeutic Clinical Trials," *Spinal Cord* 45, no. 3 (March 2007): 190–205.

200 *A team of scientists . . . surgically implanted a tiny electrical stimulator:* Susan Harkema et al., "Effect of Epidural Stimulation of the Lumbosacral Spinal Cord on Voluntary Movement, Standing, and Assisted Stepping after Motor Complete Paraplegia: A Case Study," *Lancet* 377, no. 9781 (June 2011): 1938–47.

201 *So too were three other men who were later implanted with stimulators:* Claudia A. Angeli et al., "Altering Spinal Cord Excitability Enables Voluntary Movements after Chronic Complete Paralysis in Humans," *Brain* 137, pt. 5 (May 2014): 1394–409.

202 *estimated 250,000 to 1.3 million Americans:* National Spinal Cord Injury Sta-

tistical Center, https://www.nscisc.uab.edu/Public/Facts%202016.pdf; Roni Caryn Rabin, "Study Raises Estimate of Paralyzed Americans," *New York Times*, April 20, 2009.

202 *At the University of Pittsburgh School of Medicine, scientists decided:* "Woman Guides Robot Arm with Thoughts," University Pittsburgh Newsroom, December 16, 2012, http://www.neurosurgery.pitt.edu/news/woman -guides-robot-arm; Jennifer L. Collinger et al., "High-Performance Neu-roprosthetic Control by an Individual with Tetraplegia," *Lancet* 381, no. 9866 (February 16, 2013): 557–64, https://www.ncbi.nlm.nih.gov /pmc/articles/PMC3641862/.

206 *tai chi has been proven effective for certain types of motor control:* Fuzhong Li et al., "Tai Chi and Postural Stability in Patients with Parkinson's Disease," *New England Journal of Medicine* 366, no. 6 (2012): 511–19; Lesley D. Gil-lespie et al., "Interventions for Preventing Falls in Older People Living in the Community," *Cochrane Database of Systematic Reviews* 15, no. 2 (April 2009): CD007146.

Chapter Fourteen: Believing Is Seeing

209 *"The universe is full of magical things patiently waiting":* Eden Phillpotts, *A Shadow Passes* (London: C. Palmer & Hayward, 1918).

209 *A former UCLA physical therapy professor named Valerie Hunt:* http:// valerievhunt.com/VALERIEVHUNT.COM/Valerie_Hunt_Research .html; Susan Barber, "The Promise of Bioenergy Fields: An End to All Diseases, an Interview with Dr. Valerie Hunt," *Spirit of Maat* 2 (November 2000), http://www.spiritofmaat.com/archive/nov1/vh.htm.

209 *In North Carolina, at the Rhine Research Center:* William T. Joines et al., "Electromagnetic Emission from Humans During Focused Intent," *Journal of Parapsychology* 76, no. 2 (September 2012): 275–94, http://www.rhine .org/images/jp/v76Fall2012/dJPF2012Joines.pdf; conversation with John Kruth, August 2017.

210 *Then there's William A. Tiller:* William A. Tiller and Walter E. Dibble Jr., "A Brief Introduction to Intention-Host Device Research," https://www .tillerinstitute.com/pdf/White%20Paper%20I.pdf; conversation with Bill Tiller, June 2014.

210 *a 2015 paper that outlines such a search:* Richard Hammerschlag et al., "Biofield Physiology: A Framework for an Emerging Discipline," *Global Advances in Health and Medicine* 4 (suppl.) (November 2015): 35–41, http:// www.gahmj.com/doi/10.7453/gahmj.2015.015.suppl.

213 *chiropractor/energy healer named Donny Epstein:* "Tony Robbins Talks About Network Chiropractic," https://www.youtube.com/watch?v= BX6XiMDUTdE.

213 *Chinese qigong healer who participated in studies:* Xin Yan et al., "Certain

Physical Manifestation and Effects of External Qi of Yan Xin Life Science Technology," *Journal of Scientific Exploration* 16, no. 3 (2002): 381–411, https://flowingzen.com/downloads/YanXinPhysicalManifestation.pdf; Xin Yan et al., "External Qi of Yan Xin Qigong Differentially Regulates the Akt and Extracellular Signal-Regulated Kinase Pathways and Is Cytotoxic to Cancer Cells but Not to Normal Cells," *International Journal of Biochemistry & Cell Biology* 38, no. 12 (2006): 2102–13.

215 *Quantum-Touch founder Richard Gordon insists:* https://www.quantumtouch .com/en/component/content'/?view=article%27&id=3%27&Itemid=156.

218 *Powers and . . . Corlett tested and confirmed sixteen more:* Albert R. Powers III et al., "Varieties of Voice-Hearing: Psychics and the Psychosis Continuum," *Schizophrenia Bulletin* 43, no. 1 (January 1, 2017): 84–98.

219 *estimates put it at around 10 to 15 percent:* Hélène Verdoux and Jim van Os, "Psychotic Symptoms in Non-Clinical Populations and the Continuum of Psychosis," *Schizophrenia Research* 54, no. 1–2 (March 1, 2002): 59–65.

219 *In a 2009 study at University College in London:* Oliver J. Mason and Francesca Brady, "The Psychotomimetic Effects of Short-Term Sensory Deprivation," *Journal of Nervous and Mental Disease,* October 2009.

220 *Between 30 and 50 percent of elderly widows experience:* Patricia Boksa, "On the Neurobiology of Hallucinations," *Journal of Psychiatry & Neuroscience* 34, no. 4 (July 2009): 260–62.

220 *2015 study in which 90 percent of volunteers felt moderate or strong sensations:* Florian Beissner et al., "Placebo-Induced Somatic Sensations: A Multi-Modal Study of Three Different Placebo Interventions," *PLoS One* 10, no. 4 (April 22, 2015): e0124808, https://www.ncbi.nlm.nih.gov/pmc/articles /PMC4406515/.

222 *2004 paper by Jamie Ward:* Jamie Ward, "Emotionally Mediated Synaesthesia," *Cognitive Neuropsychology* 21, no. 7 (October 2004): 761–72, http:// synaesthesia.info/Ward-04.pdf.

222 *the novelist and famous synesthete Vladimir Nabokov:* Maria Popova, "Nabokov's Synesthetic Alphabet: From the Weathered Wood of *A* to the Thundercloud of *Z*," *Brain Pickings,* https://www.brainpickings.org/2018/05/15 /nabokov-synesthesia/.

222 *a British synesthete told a flabbergasted interviewer:* Kate Samuelson, "The Man Who Tastes Sounds," *Motherboard,* April 29 2015, https://motherboard .vice.com/en_us/article/ypw4m7/the-man-who-tastes-sounds.

222 *Forty percent of synesthetes have a parent:* David Brang and V. S. Ramachandran, "Survival of the Synesthesia Gene: Why Do People Hear Colors and Taste Words?," *PLoS Biology* 9, no. 11 (November 2011): e1001205.

223 *neuroscientist V. S. Ramachandran . . . referred to it as emotion-evoked colors:* V. S. Ramachandran et al., "Colored Halos Around Faces and Emotion-Evoked Colors: A New Form of Synesthesia," *Neurocase* 18, no. 4 (2012): 352–58.

224 *blinded studies have failed to validate:* See summary of studies on applied

kinesiology in Stephan A. Schwartz et al., "A Double-Blind, Randomized Study to Assess the Validity of Applied Kinesiology (AK) as a Diagnostic Tool and as a Nonlocal Proximity Effect," *Explore (NY)* 10, no. 2 (March–April 2014): 99–108.

224 *posting on a chat board by . . . Stephen Perle:* "Applied Kinesiology and Self Deception," *Neurologica Blog,* December 7, 2007, https://theness.com/neurologicablog/index.php/applied-kinesiology-and-self-deception/.

Chapter Fifteen: Why Doctors Need to Be More Like Alternative Healers (and Vice Versa)

230 *researchers at the University of Liverpool reviewed audiotaped consultations:* Peter Salmon et al., "Voiced but Unheard Agendas: Qualitative Analysis of the Psychosocial Cues That Patients with Unexplained Symptoms Present to General Practitioners," *British Journal of General Practice* 54, no. 500 (March 2004): 171–76, https://www.ncbi.nlm.nih.gov/pmc/articles/PMC1314826/pdf/15006121.pdf.

230 *even though 60 to 80 percent of primary care visits . . . are likely:* Aditi Nerurkar et al., "When Physicians Counsel About Stress: Results of a National Study," *JAMA Internal Medicine* 173, no. 1 (January 14, 2013): 76–77, https://www.ncbi.nlm.nih.gov/pmc/articles/PMC4286362/.

232 *famous 1927 essay by the Harvard professor Francis Weld Peabody:* Francis W. Peabody, "The Care of the Patient," *JAMA* 88 (March 19, 1927): 877–82, https://depts.washington.edu/medhmc/wordpress/wp-content/uploads/Peabody.html.

233 *only 40 percent of US hospitals received a four- or five-star rating:* Sabriya Rice, "Only 251 U.S. Hospitals Receive 5-Star Rating on Patient Satisfaction," *Modern Healthcare,* April 16, 2015, http://www.modernhealthcare.com/article/20150416/NEWS/150419925.

234 *Other polls show that only 34 percent of Americans have great confidence:* Robert J. Blendon et al., "Public Trust in Physicians—U.S. Medicine in International Perspective," *New England Journal of Medicine* 371 (2014): 1570–72, https://www.nejm.org/doi/full/10.1056/NEJMp1407373#t=article.

234 *studies that evaluate medical students before and after med school:* Kathy A. Stepien and Amy Baernstein, "Educating for Empathy: A Review," *Journal of General Internal Medicine* 21 (2006): 524–30.

235 *wrote the Yale gastroenterologist Howard Spiro:* Howard Spiro, "Clinical Reflections on the Placebo Phenomenon," in *The Placebo Effect,* ed. Anne Harrington (Cambridge, MA: Harvard University Press, 1977).

Index